Quantum Legacy

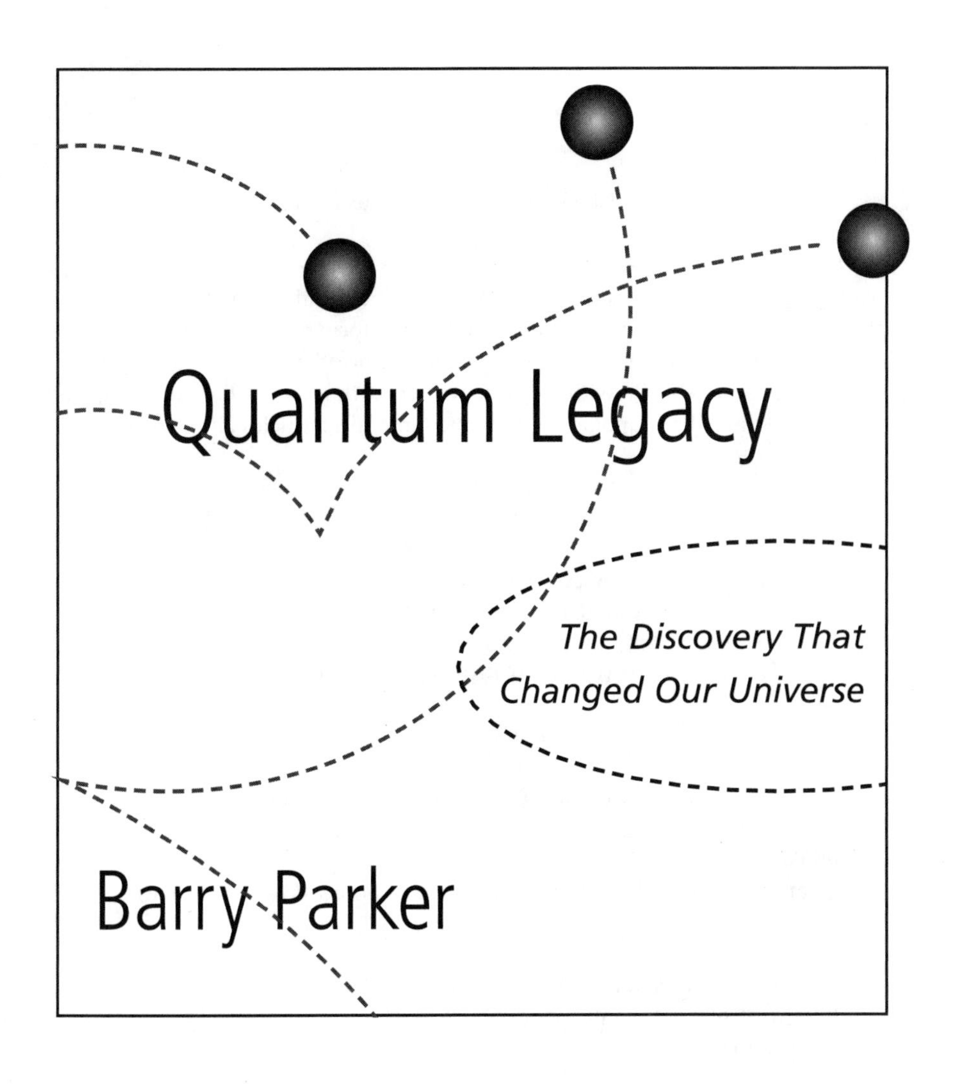

Quantum Legacy

The Discovery That Changed Our Universe

Barry Parker

Prometheus Books
59 John Glenn Drive
Amherst, New York 14228-2197

Published 2002 by Prometheus Books

Inquiries should be addressed to
Prometheus Books
59 John Glenn Drive
Amherst, New York 14228–2197
VOICE: 716–691–0133, ext. 207
FAX: 716–564–2711
WWW.PROMETHEUSBOOKS.COM

06 05 04 03 02 5 4 3 2 1

Library of Congress Cataloging-in-Publication Data

Parker, Barry R.
Quantum legacy : the discovery that changed our universe / by Barry Parker.
p. cm.
Includes bibliographical references and index.
ISBN 1–57392–993–X (alk. paper)
1. Quantum theory. I. Title.

QC174.12 .P373 2002
530.12—dc21

2002067966

Printed in Canada on acid-free paper

Contents

Preface

When I was a graduate student, one of my relatives asked me what I was studying.

"Quantum mechanics," I said.

"Quantum what?" he replied.

"Quantum theory . . . you know . . . Bohr, Schrödinger, Heisenberg."

"Never heard of them."

"What about Einstein?"

"Yeah . . . I've heard of him."

It's strange that many people have heard of Einstein and his theory of relativity, black holes, the big bang theory of the universe, and so on, but have no idea what quantum mechanics is, yet it has affected their lives much more than any of the above.

Quantum mechanics is the theory that explains the structure, motions, interactions, and general behavior of atoms and molecules. It is a complicated, mathematical theory, but we won't get into any of the mathematical details in this book.

Although the theory is complicated, it is possible for the layperson to understand many of the ideas of quantum theory. In this book I will present many of the basic ideas, but I will also tell the story of the discovery and development of the theory. It is a story that is filled with drama. Indeed, all the elements of a good fictional story are there—hardships, frustrations, hopes, surprises, and the joy associated with a great discovery. But it isn't a fictional story. It's a true one. I will introduce you to the characters of this drama and their struggles, but I will also show the importance of quantum mechanics in our world—in particular, how it has changed our world and helped make our lives more comfortable.

It is not possible to talk about science without using some scientific terms, and it is likely that you are unfamiliar with some of them. I have attempted to define each of the terms as it appears, but for the benefit of those new to science I have added a glossary at the end of the book. Very large and very small numbers are also used occasionally, and I have used scientific notation to designate them. The notation 10^{20}, for example, represents the number 1 with 20 zeros after it. Similarly, the notation 10^{-20} represents 1 divided by 10^{20}. Temperature scales are also a problem. Scientists prefer the Kelvin (K°) scale, and I have used it in several places. On this scale the lowest temperature in the universe is 0° K, which corresponds to –459° F.

The line drawings and the drawings of the scientists were done by Lori Scoffield-Beer based on illustrations in Robert L. Weber et al. *College Physics*. I would like to thank her for an excellent job and the McGraw-Hill Companies, Inc. for allowing us to use these illustrations. I would also like to thank my editor, Linda Greenspan Regan, and the staff of Prometheus Books for their help in bringing this book into its final form. Finally, I would like to thank my wife for her support while the book was being written.

Introduction

Pocket calculators, digital watches, home computers, digital radios, television, lasers, and the solid-state control devices in modern appliances and cars all depend on a branch of physics known as *quantum mechanics*. Quantum mechanics affects our lives every day, at least indirectly, yet most people have hardly heard of it. The origin of many modern devices including lasers can be traced to it. Lasers, for example, are now in grocery stores, at the doctor's office, in surveyor's equipment, in sophisticated war devices, in machine shops, in textile mills, in telephone lines, and in many other places. Furthermore, the power for your lights, heat, air conditioning, and appliances also depends, in many cases, on quantum mechanics when your power is generated by nuclear reactors. In addition, with computers now being commonplace and almost every household having a personal computer, quantum mechanics comes into play. Computers depend on integrated circuits, which in turn use transistors that have been developed because of our knowledge of quantum mechanics.

The year 2000 was the hundredth birthday of quantum theory, but there was not much in the way of celebration. In fact, I hardly remember anyone even mentioning it. The breakthrough that led to quantum theory occurred on October 7, 1900. Like most great discoveries, however, considerable time was needed for it to be appreciated. Indeed, it would be another twenty-five years before a full-fledged quantum theory would be developed.

The breakthrough, which was made by Max Planck of Germany, has been described as a "lucky guess," and even Planck would agree that it was. He was unsure he was taking the right approach; all he knew was that it worked. Strangely, it was a rather mundane problem that led to the breakthrough. You are likely familiar with the phenomenon. Heat a piece of iron and watch it change color as the temperature increases. Each color gives off a certain amount of radiation—a different amount for each color. A plot of this is easy to make, and it was this plot that physicists could not explain. Planck worked on the problem for months, and then in desperation he made a radical suggestion. He said that radiation was composed of "chunks"—what we now call *quanta*. Prior to this, radiation had always been thought of as continuous. But, to his surprise, his idea worked, and the curve was explained. Planck thought he was merely providing an interim explanation, and a better one would eventually come along. But it didn't. Unknowingly, he had opened a Pandora's box, and it would contain things that no one ever dreamed of.

Few people paid much attention to Planck's explanation at first. It was too radical, and even Planck was convinced that it was only a temporary solution. But one person was not so sure, and that person was Albert Einstein. Einstein was certain that it was the right answer, and he used it to explain another problem that had stumped physicists for several years. He postulated that light was both a wave and a particle. It was an equally strange suggestion, but again it worked.

It was in the first simple model of the atom, however, where quantum theory really came into its own. At Cambridge University, Ernest Rutherford had shown that a cloud of negatively charged electrons whirled around a small, dense nucleus of positive charge, but he had no idea what the cloud of electrons looked like, or how they assembled themselves. It was Neils Bohr of Denmark who took the decisive step. Using quantum theory, he suggested that the electrons were actually in distinct orbits, and when they jumped from one orbit to another, they emitted or absorbed radiation. With this idea he was able to explain the lines that occurred when the light from hydrogen was passed through a spectroscope.[1] His model was a first step, but the quantum theory he developed was crude and limited. It worked for only the hydrogen atom; if it was applied to even the second simplest atom, namely, helium, it gave the wrong answer. And, as for anything beyond helium, it was hopeless.

Then a young French prince, Louie de Broglie, put forward another "crazy idea." It was so strange that it didn't seem to make sense. Einstein had suggested earlier that light was both a wave and a particle. De Broglie suggested that particles of matter, such as electrons, were also both waves and particles. More explicitly, particles had a wave motion associated with them.

Einstein was again one of the few convinced that the idea had merit. To him it made sense. And within a few years it was shown to be correct. Yet, strangely the first big break in quantum theory that came after it made little use of the discovery. A young German who had just received his doctorate, Werner Heisenberg, showed that the spectrum of hydrogen and other elements could be explained using arrays of numbers. We now call these arrays *matrices*. The theory he developed was soon shown to be a major breakthrough; using it, he was able to solve many previously unsolved problems. But few people were familiar with matrices; almost no one used them in physics, and as a result, his theory was not popular.

Within a year, however, Erwin Schrödinger of Zurich, Switzerland—a man who, at forty, was considered to be well past his most creative years—hit on what appeared to be a completely different approach. His method was based on waves and it centered on a *wave equation*. This wave equation was a calculus equation referred to as a "differential equation," something all physicists used and were entirely familiar with.[2] Almost everyone liked the new approach. It made sense.

But there was a problem. There were now two theories, and they appeared to be quite different. Why two? And why were they so different? It didn't take long for Schrödinger to show that they were equivalent—just different forms of the same theory. But that didn't solve all the problems. One of the major remaining problems was the meaning of the wave function that came out of Schrödinger's equation, in other words, the solution of the equation. Schrödinger was unsure; he thought it was an electrical *charge density* associated with the electron.[3]

Then along came Max Born of Göttingen, Germany, with a new interpretation. He said it was a probability distribution, and that the atomic world was based on probabilities and statistics. You couldn't say anything certain about anything in this infinitesimal world. The clincher came with a principle formulated by Heisenberg, called the *uncertainty principle*. According to this principle, you couldn't simultaneously measure variables such as the position of an atom and its momentum (velocity) to a high degree of accuracy. If you narrowed in on one, the other became "fuzzy." There was now no doubt: the world of the atom was a fuzzy place. To top this off, Bohr suggested that objects in this world didn't become "real" until they were measured. At first this appeared to be nonsense, and this time even Einstein didn't believe it.

In fact, Einstein objected strongly to the strange idea; he presented argument after argument trying to show that this interpretation couldn't possibly be correct. But Bohr managed to counter each one. The most famous of Einstein's arguments was called the *EPR paradox*, which led to some bizarre predictions. Einstein and others thought that it was possible that there were hidden variables in the theory, in other words, a subtheory that would overcome our inability to determine things exactly. In 1965, however, John Bell of the European high-energy laboratory CERN published a relation that eventually proved that this would never be. Quantum mechanics was it, and the probability interpretation was here to stay.[4] Indeed, Bohr's idea that things in the microworld were not real until measured appeared to be right.

In its final stage of development, quantum mechanics was extended to speeds close to that of light (relativistic speeds) and to the interaction between particles and photons at high speed.[5] This extension was called *quantum electrodynamics*. Still, the strangeness did not go away. In fact, more problems were soon encountered.

The strange predictions of quantum mechanics are bizarre, and I will discuss many of them in the chapters that follow. For most people, however, it is the "spin-offs" of quantum mechanics that have affected their lives. Indeed, most people don't realize how important quantum mechanics is to our society. It has given us a "gold mine" in terms of energy-saving devices, instruments, and so on. Quantum mechanics is the basis of laser theory, solid-state physics, nuclear physics, and molecular biophysics, and the practical applications of these sciences can be seen all around us.

Lasers depend on an effect called *stimulated emission*, which was discovered by Einstein in 1916 while he was investigating quantum systems. It was known that an electron in an excited state would eventually jump down to its ground state, and in the process emit radiation. Einstein showed that the process could be hastened. In effect, it could be "stimulated," or nudged, by photons.

Stimulated emission is the central phenomenon in the laser effect. In the late 1940s, two teams, one in the United States and one in the USSR, showed that collections of atoms or molecules that had been excited could be tricked into returning to the ground state, all at the same time. The result was the *maser*, a device that used microwave radiation. Ten years later the same trick was applied to ordinary light and the *laser* was born.

The laser has now become indispensable in modern society. Some people think of "phasers" or "photon torpedoes" when they hear about lasers. They are the weapons of science fiction, and indeed a tremendous effort is going into deploying lasers for war. But lasers are used not just for war; in fact, they're used much more extensively in medicine. Eye surgeons now use them routinely for such things as cataract surgery, repairing detached retinas, and repairing broken blood vessels in the eye. Lasers are even used in the treatment of cancer. Certain dyes cause cancer cells to absorb the heat from a laser more readily than normal cells, and as a result lasers can be used to kill cancer cells while leaving normal cells unharmed. Heart surgeons also now use them routinely for removing plaque and blood clots from arteries. And the list goes on.

Solid-state physics is another branch of physics that has its origins in quantum mechanics. Indeed, without quantum mechanics it would not exist. As everyone knows, conductors play a large role in our lives—they conduct electricity to our homes. But it is materials that do not conduct so well that are the centerpiece of solid-state physics. They are called *semiconductors*. What is important about semiconductors is the distribution of electrons in their energy levels. These energy levels are so close together in a semiconductor we refer to them as *energy bands*. And quantum mechanics is needed to determine these bands accurately. The number of electrons and holes (regions where there are no electrons) in the various levels can be closely controlled, and it is this control that has made these materials so important. From them come transistors, diodes, and micro or integrated circuits, which are now indispensable to our society.

The list of things that solid-state physics has given us is too long to write down. Anything that has a transistor or integrated circuit in it owes its origin to solid-state physics and quantum theory. One of the biggest legacies of quantum mechanics, however, is the computer that now dominates our society. Personal computers are in nearly every home, allowing us to connect to the world via the Internet. Furthermore, the world of commerce and industry is now run on computers. Over the past few years, we have experienced a computer revolution. Computers have become more compact and user-friendly, so that almost anyone can use one. And of course inside a computer are thousands of transistors, diodes, and tiny electronic circuits called *chips*. Computers are now an important part of society. Most people couldn't visualize life without them, and they usually take them for granted. And quantum mechanics was instrumental in their origin.

Quantum mechanics is also at the heart of nuclear physics in that it gave us an understanding of the nucleus of the atom. The nucleus is a densely packed region of protons and neutrons, held together by a particularly strong force called the *strong nuclear force*. Like the atom, it also has energy levels that can be determined only by quantum mechanics. Without a thorough understanding of these energy levels, we would never have been able to control nuclear energy. Control of this energy has brought us nuclear reactors, and nuclear reactors supply a large fraction of the energy used on earth.

Finally, we have the science of molecular biology. As most people know, the story of the DNA molecule begins with the discovery of its structure by Francis Crick and James Watson. Crick was originally trained as a physicist and changed to molecular biology after World War II. Both he and Watson were strongly influenced by a little book written by Schrödinger titled, *What Is Life?* It is now considered to be a classic and for the first time put forward the idea that life, and the foundation molecules of life, could be understood in terms of physics and quantum mechanics.

The legacy of quantum mechanics is huge, and certainly something that is worth celebrating. Not only has quantum mechanics changed the world, but it has also changed our understanding of the universe. I will begin with a detailed look at the history of quantum mechanics in the next chapter. As you will see, this history is filled with interesting characters and events. In the latter chapters of the book, I will look in more detail at what quantum mechanics has given us.

Early Ideas

We tend to think of the atom as a relatively recent discovery, at least a twentieth-century discovery, but in reality the idea of an infinitely small unit of matter goes back to the early Greeks. The Greek philosopher Democritus, who was born about 490 B.C.E., suggested that all matter was composed of indivisible units that he called *atoms*. As you might expect, this is where our word "atom" comes from.

Democritus was known as the laughing philosopher. Although we're not sure why he acquired such a name, it's likely because he laughed so much or at least was very jovial. Anyway, according to his idea everything was composed of atoms, with atoms of different materials being different in some way. Water atoms, for example, were assumed to be round and smooth. Atoms of earth, on the other hand, were rough and jagged. Although his ideas were surprisingly modern, they were ahead of their time and few took them seriously. They were discussed for some years after his death, but were eventually overshadowed by Aristotle's idea that the basic elements of nature were air, water, earth, and fire. According to Aristotle, each of these elements had a natural place on earth, and if an element was displaced, it would return to its natural place.

DALTON

Although the idea of atoms continued to be discussed, no one had any proof or evidence that they existed. Later Galileo was convinced they existed and Isaac Newton also believed in them, but in both cases it was just intuition.

The idea was finally put on a firmer basis by the English chemist John Dalton (1766–1844). Even by the standards of the day, Dalton wasn't a highly skilled scientist, but he was competent and enjoyed experimenting. Most of all, though, he was a Quaker with a passion for weather. Each day, as regular as clockwork, he recorded the temperature, weather conditions, rainfall, and cloud cover. He did this faithfully for fifty-seven years. He kept so busy he had no time for a social life; he never even had time to marry, and remained a bachelor throughout his life.

To us he would seem like a dull person, but he was curious by nature. As he stared up at the clouds, he wondered what they were made of. What caused rain? What was air composed of? In 1800 his curiosity led him to construct a crude laboratory where he began experimenting with air and other gases such as hydrogen and oxygen. He soon showed that oxygen was exactly eight times as heavy as hydrogen, and he later discovered the law of partial pressures. It seemed reasonable to him that each gas was made up of atoms, with atoms of different gases having different masses. He soon went beyond gases and assumed that all elements were composed of atoms, with the only real difference between atoms of different elements being their masses. Furthermore, he distinguished between "elements" and "compounds," suggesting that compounds are composed of two different kinds of atoms.

Dalton published his ideas in 1802 and 1808 under the title *New*

System of Chemical Philosophy. The idea made so much sense and seemed to explain so many things, it was soon accepted by most scientists. But not everyone liked it. To some it was repugnant, and they criticized it, but the criticism didn't bother Dalton. There was overwhelming support for his theory, and he was confident it was correct. His supporters showered honors on him, but he didn't like the honors much better than the criticism. He shunned publicity and disliked too much attention.

Oxford University bestowed an honorary doctorate on Dalton in 1832, but when he was told he would have to wear scarlet robes to accept the degree, he was startled. Scarlet was a forbidden color for Quakers. The problem was solved, however, when officials realized he was color-blind. To him the robes appeared gray.

Upon his death in 1844, he was given an elaborate last tribute. His funeral was attended by thousands.

THE KINETIC THEORY OF GASES

Within a few years, the idea of an elementary, indivisible particle had been accepted by most scientists. There was a problem, of course: these particles had never been seen. Indeed, although their size was unknown, it was assumed that they were far beyond the range of the best microscopes. Nevertheless, there seemed to be a tremendous amount of evidence for their existence. Robert Hooke had put forward the idea that the pressure exerted by a gas in a vessel was due to tiny particles colliding with the walls of the vessel. These particles had to be the molecules of Dalton. When the gas was heated, the molecules moved faster and exerted more pressure. Furthermore, something particularly interesting had been discovered. If a plot of pressure was made against temperature for several gases, and the plot was extrapolated to low temperatures, the lines all converged to the temperature –273° C (–459° F). What was the significance of this temperature? It seemed reasonable to assume it was the temperature at which the molecules of the gas were at rest. In other words, it was the lowest possible temperature: "absolute" zero. This solidified the idea that molecular motion was a measure of heat, and in the mid-1800s Rudolf Clausius of Germany showed that the temperature of a gas was a measure of the kinetic energy of the molecules. At higher temperatures molecules moved at higher speeds and had higher energies.

But if the molecules of a gas were like little billiard balls, Newton's laws of motion could be applied to them, and many things could be calculated. Actually, because there were so many particles, and they were all identical, statistics (later called statistical mechanics) had to be used. Nevertheless, many important predictions were made. Eventually a new theory, which became known as the *kinetic theory of gases*, was developed. The two names most closely associated with it are Ludwig Boltzmann and James Clerk Maxwell. Maxwell is best known for his development of electromagnetic theory.

In 1911 Amedeo Avogadro showed that at a fixed temperature and pressure, equal volumes of a gas contained the same number of molecules; in other words, one *mole* (one molecular weight) of a gas contained approximately 6.023×10^{23} molecules.[1]

SIZE OF THE ATOM

Even though the idea of atoms and molecules was accepted by most people, several well-known scientists did not believe it. Something more was obviously needed.

This was the situation when Albert Einstein began searching for a topic for his doctoral thesis. One day while talking to his friend Michele Besso, Einstein poured sugar into his tea and began stirring it. As he stirred, he thought about the tea. "When I put sugar in the tea, the viscosity is changed," he said to Besso. Suddenly he realized that the viscosity would depend on the size of the sugar molecule. Making some rough calculations, he was able to show that the change in viscosity was, indeed, related to the volume of the sugar molecule.

That evening he returned to the problem. He soon had what he wanted: an expression for the size of the sugar molecule. As he had anticipated, it depended on both the viscosity and the *diffusion coefficient* (a measure of the rate at which sugar diffuses through a porous medium) of sugar. The following day he looked up the data he needed for sugar and found that the sugar molecule had a diameter of one ten-millionth of a centimeter. He wrote it up and used it as his thesis.[2]

Despite the important result it contained, the thesis was returned to him soon after he submitted it. His thesis advisor, Alfred Kleiner, thought it was too short (it was seventeen pages long). Einstein went through the thesis carefully, added one sentence to it, and sent it back to Kleiner. Surprisingly, Kleiner now accepted it.

But Einstein wasn't finished. He wanted an even more direct way of determining the size of an atom or molecule. In his thesis he had considered a sugar molecule; he now realized that if water molecules knocked a sugar molecule around they would also knock very small objects around. These objects could be much larger than molecules, however, and they might be large enough to be seen under a microscope.

He told Besso about his idea, and Besso informed him that he was talking about *Brownian movement*. Einstein had never heard of the effect. Besso explained that about seventy-five years earlier the botanist Robert Brown had observed tiny pollen grains in water using a microscope and had noticed a strange trembling motion. He later showed the same thing occurs for finely ground splinters of glass and other materials.

Einstein began thinking about the problem. Could he use it to determine the size of the molecule? The answer came to him the following Sunday when he was caring for his young son. He thought of the large spheres of pollen being knocked around by the vibrating molecules of water. He knew there was no way he could calculate their speed directly; a simple calculation had shown him that a sphere of pollen would move over a hundred times its diameter every second if unhindered. But it was hindered by the water molecules and would undergo collisions, so its overall movement would be greatly changed. He couldn't calculate the displacement of a single sphere, but he could calculate the *average displacement* of one, and he was sure this would be measurable under a microscope. He made the calculation and determined that a tiny sphere with a diameter of one thousandth of a millimeter would move by one thousandth of a millimeter in one second, or sixty thousandths of a millimeter in a minute. This would be observable.

Over the next few days, he wrote up the paper and sent it in for publication. This paper would eventually be considered a classic. He didn't prove that atoms existed, but he did predict exactly how far molecular collisions would push a small grain of pollen or other material, and observations soon verified his prediction.

CATHODE RAYS

The atom, by definition, was the ultimate elementary particle of nature. But was it structureless? Perhaps it was made up of more elementary particles. What did it look like up close? Furthermore, there was a problem: many people believed there was another tiny particle associated with electrical current. If so, how was it related to the atom? And Michael Faraday had shown that there were positively charged ions. Where did they fit in?

Part of the answer was to come from an apparatus that was used extensively at the time. It was called a *cathode ray tube*. Electrical current could actually be seen in cathode ray tubes, but this current was referred to as cathode rays. The apparatus was an evacuated tube with two terminals (metallic caps) implanted in it that could be attached to the terminals of a power source. The leads were sealed in the glass. If a potential difference of about 10,000 volts was applied and the air was pumped out of the tube, a fuzzy spark appeared down the center of it. As the tube was evacuated further a glow appeared, and interestingly, the glow changed color when different gases were introduced into the tube.

The negative terminal of the tube was referred to as the *cathode*, the positive terminal as the *anode*. And if you cut a small hole in the anode, the beam would continue on through it and strike the glass at the far end of the tube causing it to glow, or fluoresce.

What was the beam composed of? Some thought it was a new form of light; others were convinced that it was a beam of particles. But there were problems. If they were particles, they appeared to be quite different from Faraday's ions. First of all, they had the same properties regardless of what gas was used to fill the tube. Second, they easily passed through thin films of matter. Faraday's ions did neither.

Although little was known about these rays, and atoms in general, the outlook within the physics community was gloomy. There was considerable apathy about the future. Many scientists were beginning to believe that almost everything in physics had been discovered, and its glory days were over. Students were beginning to be discouraged, and few were selecting physics as a career. All this changed suddenly in 1895 with a significant discovery. It told us little about atoms or electrical currents. Indeed, in many ways, it confused things. Nevertheless, it was a great discovery, and the next few years would be some of the most fruitful in the history of physics.

THE DISCOVERY OF X RAYS

The discovery that amazed everyone, even the general public, was a strange new type of radiation. For lack of a name, the discoverer, Wilhelm Röntgen, called it x rays. The letter x is usually the unknown in an algebraic equation, and these rays were an "unknown."

Wilhelm Konrad Röntgen was born on March 27, 1845, in Prussia. He obtained his doctorate at the University of Zurich. When he made his discovery, he was head of the Department of Physics at the University of Würzburg in Bavaria. He was fifty years old at the time of the discovery, not in his prime, and like many physicists he was working on cathode rays. He was particularly interested in the luminescence that these rays caused in certain chemicals. It was so faint that it could be seen only in a dark-

Wilhelm Röntgen

ened room. Furthermore, because the cathode ray tube gave off a glow, he had surrounded it with black paper so that he could observe the luminescence better. He was observing the effect in November 1895 when a flash of light across the room caught his attention. Investigating further, he found the light came from a sheet of paper that was coated with the luminescent material barium platinocyanide.

He was puzzled. The cathode rays could not be causing the luminescence. The light from them was blocked off by the black paper around the tube. Yet when he turned the tube off, the luminescence in the sheet disappeared. It was there only when the tube was turned on.

What was going on? There had to be some other radiation emanating from the cathode ray tube. It was obviously highly penetrating, but invisible to the eye. He continued experimenting, finding that the rays could penetrate wood and even thin sheets of metal. He also soon discovered that the rays were coming from the place where the cathode beam struck the glass. There was considerable luminescence at this point. Over the next few weeks, he worked feverishly to determine as many of the properties of the rays as he could. He found that they could ionize gases (create charged particles in them), but they did not respond to electric and magnetic fields.[3]

On June 23, 1895, Röntgen gave his first public lecture on the new rays. By now he had determined that you could see the bones of your hand using them and realized they would be important from a medical standpoint. At the end of the lecture, he took an x-ray photograph of the hand of one of the men in the audience. The audience was amazed. Röntgen published an account of the discovery in the proceedings of the local scientific society. A translation appeared in the British journal *Nature* on January 23, 1896. A photo of the bones of a hand accompanied the article. It amazed both scientists and the general public, and within a short time x rays were being used to locate bullets and other foreign metallic objects in the body.

Röntgen was awarded the first Nobel Prize in 1901 for his discovery. Within a year of the discovery, over one thousand papers were published about it. Röntgen's x rays were eventually shown to be electromagnetic waves. Many considered the discovery as the beginning of modern physics.

RADIOACTIVITY

The discovery of x rays was followed almost immediately by a discovery that would be more important in relation to the structure of the atom. Indeed, it would be the key to many important discoveries that would occur over the next few years. The discovery was made by Antoine Henri Becquerel of Paris. Becquerel started out as an engineer but got his doctorate in 1888 in physics. It was perhaps inevitable that he would switch to physics since both his father and his grandfather were physicists. His father, in fact, had made several important discoveries.

In 1891 Becquerel took the same position at the Museum of Natural History in Paris that his father and grandfather had held. He even followed in their footsteps by continuing their research in fluorescence and phosphorescence. When the discovery of x rays was announced in January of 1896, Becquerel was intrigued. With his interest in fluorescence, he wondered if any of the fluorescing materials he had been working on emitted x rays. After all, the cathode ray tube fluoresced where the x rays were generated.

Within a few weeks, Becquerel began his experiments. He wrapped photographic paper in heavy black paper and put fluorescing chemicals on top of it. He then took it out into the sun, assuming the sun would induce fluorescence in the chemicals. If x rays were generated in the fluorescence, they would penetrate the black paper, but light could not get through it. When he developed the photographic paper, he would be able to see if x rays were present. To his delight they were. He saw the image of the chemicals on the photographic plate clearly. Eager to try the experiment again, he was disappointed to find that the sky was overcast. He waited for the next day, but it was also overcast, so he put the paper with the chemicals on the top away in a drawer.

He continued to wait for a sunny day, and by March 1 he was exasperated and frustrated. He looked at the wrapped photographic plate with the chemicals on it in the drawer. The chemical, incidentally, was a phosphorescent salt of uranium. He had nothing to do, so he decided to develop the photographic plate. After so many days, there might be a small amount of fogging of the plate due to earlier luminescence.

To his surprise the plate was fogged as much as it had been earlier. The sun obviously was not needed for the radiation; the chemicals did not need to luminesce. He could hardly believe the result. To convince himself, he placed the same chemicals on the photographic plate in a darkened room and left them for a while. And when he developed the plates, they were fogged in the same way. The radiation had to be coming from the chemicals. It was like x rays in that it penetrated in the same way; furthermore, he soon showed that it ionized gases.

The announcement of the discovery caused another sensation. Scientists throughout Europe began experimenting with the new phenomenon. Becquerel soon made several important discoveries about the radiation. He noted that it could be deflected by a magnetic field, so at least part of it was charged particles. He later showed that these particles were electrons.[4]

Pierre and Marie Curie soon became experts in the area and also made several important discoveries. Although the radiation was first referred to as Becquerel rays, Marie Curie later named it *radioactivity*, and the name stuck.

Becquerel was awarded the Nobel Prize for his discovery in 1903.

THE ELECTRON

After years of apathy about the future of physics, two of the most important scientific discoveries ever made opened physicists' eyes. Indeed, scientists had barely got over the surprise of radioactivity when J. J. Thomson announced another important breakthrough.

Born near Manchester, England, in 1856, Joseph John Thomson entered Cambridge University in 1876. In 1884 he assumed a professorship and was made director of the Cavendish Laboratory. Like many others at the time, Thomson became interested in cathode rays. It was known that they were negatively charged and appeared to be particles, but there was no proof. The major difficulty was that, although they were easily deflected by magnetic fields, they did not appear to be affected by electric fields, and if they were particles, they should have been.

Using highly evacuated tubes and exceedingly high electric fields, Thomson was finally able to show in 1897 that they were, indeed, deflected by electric fields. The reason why it was so much harder to deflect cathode rays with electric fields, as compared to magnetic fields, was that the force on a charged particle in a magnetic field depended on the particle speed along with its charge. The force on it in an electric field depended only on the charge.

J. J. Thomson

Comparing the deflection in electric and magnetic fields, Thomson was able to calculate the velocity of the particle. It was exceedingly high—more than one-tenth the speed of light. Nothing had ever been found in nature that moved this fast. With the speed known, Thomson could use the amount of the deflection to calculate the acceleration of the particle. This, in turn, gave the ratio of charge to mass (e/m, where e is charge and m is mass). The result was a complete surprise. It was 770 times greater than that for the charged hydrogen ion.

What did this mean? There were two possibilities: either the charge of the particles was much greater than that of the hydrogen ion, or the mass of the particle was much smaller. It was soon shown that the charges were the same and the mass was smaller. Thomson's value, incidentally, was only an approximation; we know that the actual difference in mass between the hydrogen ion (a proton) and the new particle (an electron) is 1836.

The new particles were soon accepted as the units of electrical current. Earlier, the Irish physicist, George Stoney, had suggested the name *electron* for the unit charge of the ion. The German physicist H. A. Lorentz began referring to Thomson's particle as the electron (over Thomson's objection), and the name stuck.

Thomson soon realized the electron had to be part of the atom. There were, after all, atoms of gas in the tube, and the electrons had to come from them. He, therefore, incorporated the electron in his model of the atom, which he presented in the early 1900s. It consisted of a positively charged background with negatively charged electrons embedded in it. Thomson referred to it as the "plum pudding" model.

Thomson was awarded the Nobel Prize in 1906 for his work on the electron. After 1906 he became interested in the stream of positive ions that were being generated in vacuum tubes. They were referred to as *channel rays*. Thomson was able to deflect them in both electric and magnetic fields as he had done with cathode rays. In 1912, while deflecting the positive ions from a tube of neon, he noticed that they deflected to two different positions. Neon, apparently, had two *isotopes*—elements with the same number of protons, but different numbers of neutrons (the other particle in the nucleus of the atom).

SPECTRA

If progress was to be made in proving or disproving Thomson's model, something else was needed. Interestingly, it was already available. Since Newton's time, it had been known that if the light from a heated metal was passed through a prism you would get a rainbow of colors. Since then, it had been determined that the light from a heated gas, when passed through a slit and then through a prism, gave strikingly colored lines. For a given gas, a few sharply defined lines appeared, and they were always in the same place, but their position differed for different gases. They were referred to as the *line spectrum* of the gas.

But Maxwell had shown that electromagnetic waves were produced when a charge oscillated, and light was an electromagnetic wave. Light of different colors therefore had different vibrational frequencies, and this meant that vibrating electrons should give off radiation, and the frequency of this radiation would be helpful in verifying or disproving Thomson's model. The electrons within the background charge could vibrate and give off radiation. In particular, if they were disturbed by a collision or some other outside influence, they would oscillate. The size of the atom was known to be about 10^{-8} cm, and with the approximate charge of the electron known, calculation of the expected radiation could be made. The emitted radiation had a frequency very close to that expected from the calculations. This was encouraging to Thomson and his colleagues.

There was, however, another model of the atom: the planetary model. In this model the electrons were assumed to be orbiting the center of the atom. And an orbiting charge would radiate in the same way a vibrating charge would. Indeed, with an orbital diameter of approximately 10^{-8} cm, it would radiate in the correct range. This was encouraging, but there was a problem. There was no way to stop the radiation; the atom would radiate continuously, and as it radiated it would lose energy. In fact, it was soon shown that the electron would spiral into the center as it lost energy, and it would take only a millionth of a second to do this. Interest in the planetary model was, therefore, short-lived.

Most scientists were convinced there was a close link between the structure of the atom and line spectra. Over the years several things had been learned about spectra. Walther Ritz of Germany showed that all spectral lines had frequencies that were either the sum or the difference of the frequencies of the other lines. A Swiss schoolteacher, Johann Balmer, was able to show that the lines of hydrogen obeyed a simple mathematical relationship involving integers. And a Swedish scientist, Johannes Rydberg, later showed that the same formula held for all elements close to hydrogen in the periodic table.

RUTHERFORD

Thomson's model of the atom was accepted by most scientists, but there were problems. Within a few years it would be completely overthrown. The man who would overthrow it, Ernest Rutherford, was born on a farm in New Zealand in August 1871. The nearest town was Nelson, which had a population of about 5,500. Rutherford attended school in Nelson, and showed considerable promise, even from an early age. Upon graduation he attended Nelson College, and after three years he obtained a scholarship to Canterbury College in Christchurch. He took a bachelor's degree in 1893. While at Canterbury College, he developed an interest in a discovery that had been made a few years earlier by the German physicist Heinrich Hertz. Maxwell had predicted the existence of electromagnetic waves in the mid-1800s, and Hertz had shown that they exist. It was an important discovery, and Rutherford was intrigued with it. Using crude apparatus, he was able to generate and receive electromagnetic waves, and he continued experimenting with the waves after he graduated. He managed to transmit Hertzian waves from one end of a shed to the other. In 1893 he published a paper on his results.

He continued experimenting but realized that he would have to leave New Zealand if he was to get anywhere in his chosen field of physics. He applied for a scholarship to Cambridge University in England in 1845, but he placed second and the scholarship was

Ernest Rutherford

offered to a chemist. Disappointed, he returned to his parents' farm in the country. He was hoeing potatoes when his mother brought him a letter telling him that the winner of the scholarship had declined it, and it would now be offered to him. He flung down the hoe and yelled, "That's the last potato I'll ever dig." And it probably was.[5]

Rutherford elected to work in J. J. Thomson's laboratory, and late in 1895 he arrived in England. Thomson wrote to him in London, welcoming him to Cambridge. Rutherford decided to continue his New Zealand research on Hertzian waves. Within a short time he had transmitted the waves across a distance of half a mile; a little later he detected signals two miles away from the source. This was a record; Guglielmo Marconi had barely begun his work and was still well behind Rutherford. In 1896 Rutherford presented a paper titled "A Magnetic Detector of Electromagnetic Waves and Some of Its Applications" to the Royal Society. Thomson was impressed.

Rutherford had only been at the lab for a year when Wilhelm Röntgen made his discovery of x rays. Thomson was fascinated by the discovery and encouraged Rutherford to look into it. Rutherford was reluctant because he wanted to continue his work with electromagnetic waves, but he soon realized the importance of the new discovery. Along with Thomson, he began investigating the ionization of gases by x rays. Both positively and negatively charged ions were produced when the molecules of a gas were bombarded with x rays. Rutherford and Thomson used magnetic and electric fields to investigate their properties.

As we saw earlier, Becquerel discovered radioactivity the following year, a discovery that was to have an even greater impact on Rutherford. He would spend the next decade working almost entirely on radioactivity. By 1898, after only three years in Thomson's lab, Rutherford had already gained a reputation, and several universities were after him. He accepted an offer from McGill University in Montreal, Canada.

While still in England, Rutherford had determined that there were two types of "radiation" given off by uranium. He referred to them as alpha and beta rays. The beta rays were later shown to be high-speed electrons, and the alpha rays were ionized helium atoms. One of Rutherford's first discoveries at McGill was a new class of radioactive substances. In collaboration with Fredrick Soddy, Rutherford found that uranium formed a different substance when it gave off radiation. The radioactive element thorium did the same thing. Both substances broke down as a result of their radioactivity into a series of intermediate elements. Furthermore, the intermediate elements continued to break down to new elements. Rutherford noticed that half the material for a given element was gone in a certain fixed time; he referred to it as the *half-life* of the element. Uranium, for example, had a half-life of 4.5 billion years; radium had a half-life of 1,620 years and thorium 234 (234 is the *mass number*, which is equal to the number of protons and neutrons) had a half-life of 24.1 days. Of particular importance, though, Rutherford showed that the radioactive elements were undergoing spontaneous transformation. They were changing from one element to another—the age-old dream of the medieval alchemists.

In 1906 Rutherford was offered the chair of physics at Manchester University in England. He had longed to get back to England, which he had always thought of as the center of activity in physics research, so he accepted the offer. Just before he left, he had begun experimenting with alpha particles. He had projected alphas at thin sheets of the clear, transparent mineral referred to as *mica* and had noticed that they underwent a small amount of scattering—of the order of one or two degrees. Most of the alpha particles passed through unaffected, and it was only a small fraction that were actually scattered. But these few made him wonder. In 1908 he was awarded the Nobel Prize in chemistry for his work on the nuclear atom and on radioactivity. He was surprised because he had never thought of himself as a chemist, and he had always had a poor opinion of chemists.

One of the first things Rutherford looked for when he got settled at Manchester was a technique that would enable him to detect individual alpha particles. Within a short time, he had two methods. One was simply to observe them striking a fluorescent screen using a microscope. Tiny flashes, referred to as *scintillations*, could be seen if your eyes were adapted to the darkness.

In 1907, his first term at Manchester, Rutherford made up a list of research projects. One of them was the scattering of alpha particles by thin metallic sheets. The experiments he had done at McGill had convinced him that the atom was mostly empty space. Most of the alpha particles were undeflected, and it was easy to show that this indicated an atom consisted mostly of empty space. He encouraged his assistant, Hans Geiger, to continue the experiments.

Shortly after Geiger got started, he came to Rutherford. "Don't you think young Marsden should have a research project," he said. Ernest Marsden was a twenty-year-old undergraduate who had shown considerable promise. Rutherford thought about it for a moment, then said, "Let him see if any of the alphas are scattered through large angles."[6]

The experiment was set up (fig.1). Alpha particles were projected at a thin sheet of gold foil. It was so thin (fifty thousandths of an inch) that there were only about two thousand atoms along the route of the particle as it passed through. Geiger was sure the experiment would be a waste of time. It would be like firing a cannon ball at a swarm of bees. As expected, most of the alphas went directly though the gold foil, with only a slight deflection. Marsden, however, was looking for large deflections, and to everyone's surprise he found some. Not only were a few scattered at 90 degrees, but he found a few that came flying back at him. They were being deflected by 180 degrees.

Geiger went to Rutherford's office. "We have seen some alphas

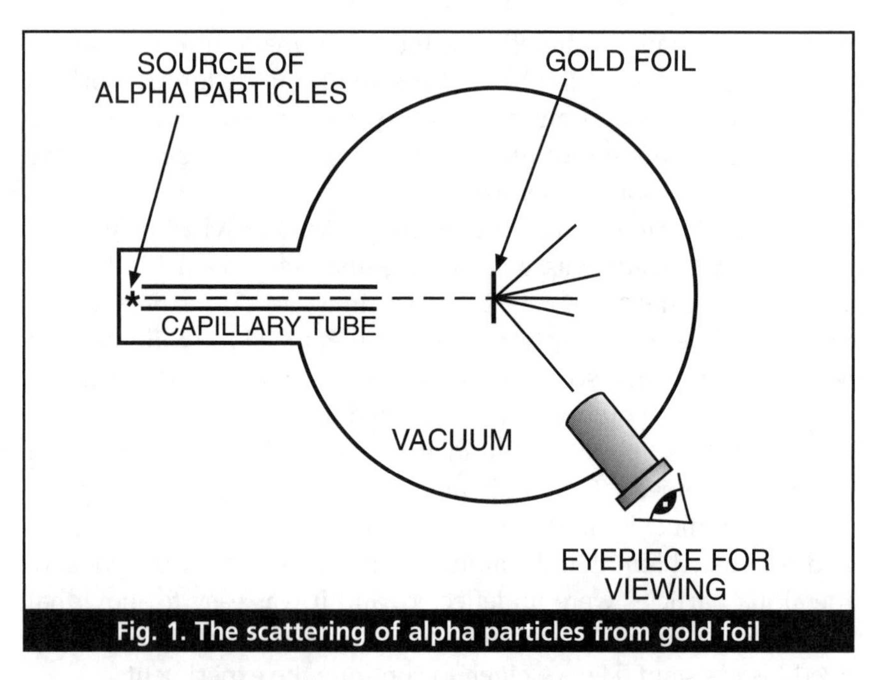

Fig. 1. The scattering of alpha particles from gold foil

that are being deflected backwards," he reported. Rutherford was stunned; he couldn't believe it. It seemed impossible. But indeed it was true. It was as if a tiny fly had deflected a massive, high-speed shell. The experiment was completed in 1909 with Rutherford still in a state of confusion. For over a year, he pondered the result. Finally, in early 1911 he announced to Geiger that he had determined what was going on. "Each of the large-angle deflections has to be due to a single encounter, and the only single collision that could produce such a deflection would be one with a very small, heavy charged particle," he said. At the center of the atom, there had to be what he referred to as a "nucleus." Virtually all the mass of the atom was in this nucleus. The model was similar to the solar system (where most of the mass is in the Sun) and was frequently referred to as the *solar system model*.

What about the electrons? They were indeed within the atom, perhaps in a cloud around the nucleus. Rutherford did not speculate on this. He decided to leave the details to the theorists.

The Lucky Guess

One of the major problems in physics in the late 1800s was explaining how and why radiant energy was emitted from a heated, glowing metal. It was known that when a metal was heated it gets red, then yellow, blue, and finally bluish white. The change in color was due to a change in the frequencies of light that were emitted. Different frequencies meant different colors, and since frequency was associated with wavelength, different colors also had different wavelengths.

BLACKBODY RADIATION

Gustav Kirchhoff of the University of Heidelberg was the first to formulate and discuss the problem. He realized that the radiation output depended on the radiative properties of the metal that was heated. To get around this, he suggested that scientists direct their attention to a "perfect" absorber and emitter, where the radiation would be independent of the material making up the body. He referred to this object as a *blackbody*. As it turned out, it was an unfortunate name. A body doesn't have to be black to be a perfect absorber and emitter. Our Sun, for example, is very close to a blackbody, and it isn't black.

Kirchhoff was born in Königsberg, Prussia, on March 12, 1824. He obtained a doctorate from the University of Königsberg in 1847 and took a position at the University of Heidelberg in 1854. Later he went to the University of Berlin. For many years his major interest was the spectra of metals. The three basic laws of spectra are, in fact, named after him. With his interest in spectra, it is perhaps natural that he would develop an interest in heat radiation.

Kirchhoff pictured his idealized, perfect radiator as a closed container with black inner walls. The only opening to the container would be a tiny pinhole. Any radiation entering the hole would have a low probability of finding its way out, and would, in effect, be absorbed. To a good approximation, it would be a perfect absorber and emitter. Furthermore, if the material within the tiny cavity was heated to incandescence, radiation would emerge from the hole and could be studied. What was important to Kirchhoff was the spectrum of the radiation, in other words, how much radiation was emitted at each frequency (or wavelength). We refer to this as its *intensity*.

Kirchhoff suggested that an accurate plot of intensity versus frequency be made and that theoreticians account for the curve by giving a mathematical formula for it. Neither of these were, however, possible in his time. It wasn't until after his death in 1887 that technical skills developed to the state where an accurate plot could be made. And the first attempts at fitting the curve with a mathematical formula didn't come until near the end of the century.

When the first plots were made, it was noticed that the major characteristic of the curve was that it was peaked. In other words, it was bell-shaped, with little radiation given off at high and low frequencies. Furthermore, the peak shifted as the temperature changed. This was seen in the changing color of the radiation. Most of the radiation was given off in the region of the peak, so the

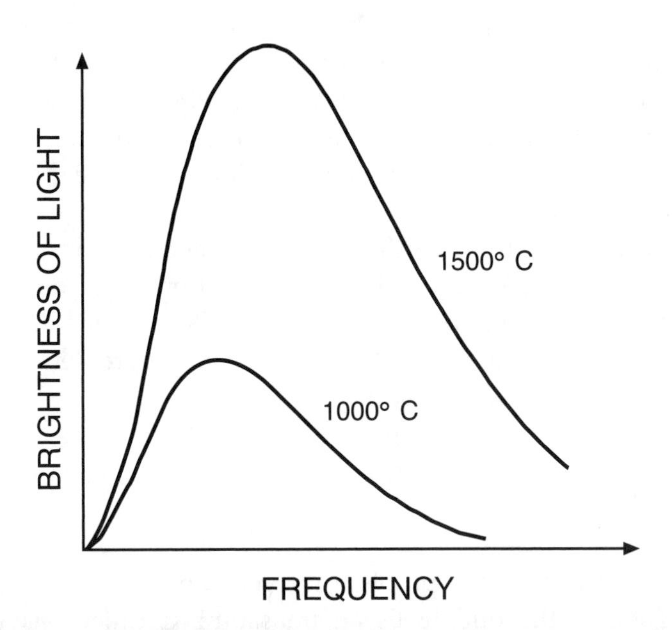

Fig. 2. Blackbody curves

color of the glow corresponded to the frequency here. As the temperature increased, the peak moved to higher frequencies and the color changed.

Lord Rayleigh of England gave the first formula for the curve, but it was valid only for low frequencies; it failed at high frequencies, in particular those corresponding to violet and ultraviolet light. Wilhelm Wien of Germany worked out another formula, but it fit the curve only at high frequencies and failed at low frequencies. It was now becoming increasingly obvious there was a problem.

To understand the problem, let's begin with a simple string. We know that a stretched string of a given length can vibrate with two, three, or more loops along its length if it is moved in the right way. Indeed, the number of modes of vibration, or number of ways the string can vibrate are, in theory, unlimited. This means that the number of different modes in the high-frequency range is going to be much greater than the number in the low-frequency range. This is like comparing the number of integers above one thousand to the number below; there are obviously many more numbers above.

Rayleigh pointed out, on the basis of this, assuming that a blackbody radiated all frequencies of electromagnetic waves equally, that the contribution from the high-frequency range would completely overshadow that from the low-frequency range. In other words, the intensity of the radiation should be greater the higher the frequency.

Experimentally this wasn't born out. In fact, there was a cutoff above certain frequencies, where no radiation whatsoever was given off. This didn't appear to make sense. Because the major difficulties were associated with the ultraviolet, the problem became known as the *ultraviolet catastrophe.*

THE LUCKY GUESS

Scientists worked on the problem for years, but no progress was made. Then in the mid-1890s an unassuming, quiet, easygoing physicist by the name of Max Planck became interested in the problem. Planck was born on April 23, 1858. When he was nine, his family moved to Munich, Germany, where he received his early education. Upon graduating from the German gymnasium, he went to the University of Berlin where he studied under Kirchhoff and other well-known scientists of the day, including Hermann Helmholtz and Rudolf Clausius. He received his doctorate in 1879, and in 1880 he joined the faculty of the University of Munich. In 1889 he succeeded Kirchhoff at the University of Berlin.

Planck's doctoral thesis was on *thermodynamics*—the study of heat and the transfer of heat between bodies—and his interest in the area continued over the following years. With Kirchhoff being one of his teachers, it was perhaps inevitable that he would become interested in blackbody radiation and the radiation curve. He no doubt heard of the difficulties of trying to fit it with a theoretical curve during his student years, but it wasn't until later that he developed a serious interest in it.

He began working on the problem in 1894. He was not looking for a revolution in physics, and, indeed, he was far from a revolu-

Max Planck

tionary. Moreover, to most he did not appear to be overly brilliant; he was extremely competent, but not a genius. In school he had never ranked at the top, rather, he usually ranked third or fourth. Most people who knew him said that his success in physics came from long, detailed study with dedication and concentration. Once Planck understood something, he understood it deeply. He knew its implications, and he had tremendous insight into what it implied.

Planck worked for several years on the problem of radiant energy with little success. But by the late 1890s he had a deep understanding of it. He was helped by the fact that several experimentalists at the University of Berlin and nearby institutes had taken an interest in the problem. Otto Lummer and Ernst Pringshein of the Physical-Technical Reich Institute in Berlin had developed several important techniques, and Heinrich Rubens and Ferdinand Kurlbaum had made considerable progress in plotting an accurate blackbody curve.

It was Sunday, October 7, 1900, when Planck finally made his breakthrough. He was forty-two at the time. Rubens and his family were visiting the Plancks for dinner. With an experimentalist and a theoretician who were working on the same problem spending an evening together, the talk eventually got around to their work. Rubens told Planck that his latest measurements had shown that at very long wavelengths (low frequencies) the energy density of blackbody radiation was proportional to temperature. In other words, at high temperatures the energy density was higher. Planck was fascinated by the information, and it inspired him to think more deeply about the difficulties he had encountered in trying to fit the curve.[1]

Later that evening when the guests had left, Planck went to his study and took another stab at the problem. Perhaps his approach had been wrong, he thought. He knew that most difficult math problems were solved when a guess is made at the answer, then checked. He went through a process of trial and error. What was needed to give a curve that folded over, like an inverted U, with low emission at high and low frequencies? Rayleigh and Wien had failed in their attempts. What had they done wrong? Finally, late in the evening he got what he wanted: a formula that gave a bell-shaped curve similar to the well-known experimental curve. It had two unknown constants in it, but they were not a problem; their values could easily be determined.

Within days he reported his success to Rubens and his colleagues, and they checked it out. To their amazement it worked. It fitted the experimental curve beautifully, and they reported the result to Planck. Planck was pleased, and on October 3, 1900, he reported the results to the German Physical Society. It is almost certain that no one at the meeting realized the significance of the relation he presented. To them it was no doubt just the result of some mathematical "juggling" that had enabled him to fit the experimental curve. It wasn't until later that it was realized that this was the beginning of the end of classical physics.

Planck was pleased with his formula, but he knew that many problems remained. He had literally guessed at the formula; indeed, he later referred to it as a "lucky guess." In all fairness, however, it can't be considered a "shot in the dark." Planck had worked at the problems for years and had tremendous insight into it, so he knew what was required. The problem at this stage, however, was that there was no physical basis for his formula. He had to explain why it worked and why it was valid. Over the next six weeks, he worked feverishly on this aspect of the problem. He later referred to this as the hardest, most strenuous work of his life.

Of particular interest were the two constants. One, which we usually designate as k, was found to have the value 1.38×10^{-23}

joules/°K and is an important constant in thermodynamics. It was named for Ludwig Boltzman, even though Boltzman had little to do with its discovery. He did, however, do important work related to heat and thermodynamics. The other constant, which we now refer to as h, was of more immediate interest to Planck. It was intricately related to the interpretation of the formula that he had come up with. Planck saw that he had to assume the radiation was emitted and absorbed in "clumps" or *quanta* as he called them. This didn't seem reasonable to him at first. Ever since Maxwell had formulated his theory of electromagnetism, it had been assumed that electromagnetic radiation was a wave, and as such was continuous. Furthermore, Hertz had detected electromagnetic waves, and there didn't appear to be any discreteness associated with them.

Planck, however, found that he was forced to interpret his result as a discreteness of the radiation. He found that energy was related to frequency; in particular, the proportionality constant was his constant h and it had a value of 6.55×10^{-12} erg sec. It was extremely small, but it wasn't infinitesimally small, as it would have to have been for the radiation to be continuous.

Planck was not completely satisfied with his explanation, but he thought a more complete understanding would eventually come and his constant would somehow eventually be incorporated into "mainstream" physics. To us, his "mainstream" physics was classical physics—the physical theories of Newton, Maxwell, and so on that existed before quantum theory was formulated. Planck certainly wasn't aware of the radical nature of his work and remained convinced for some time that he was merely applying a mathematical trick. If he had realized he was shaking up the foundation of physics, he likely would have been much more hesitant.

On December 14, 1900, he presented his interpretation of the radiation formula to a meeting of the German Physical Society. Most physicists now look on this as being the beginning of the era of quantum physics. The name "quantum physics" comes from Planck's units of energy, which he referred to as *quanta*.[2]

His pronouncement, "We therefore regard—and this is the most essential point of the entire calculation—energy to be composed of a very definite number of equal finite packages, making use . . . of the constant *h*," is certainly one of the most revolutionary statements in the history of physics. But strangely, it was hardly mentioned in the proceedings of the meeting.

It was now easy to see why the radiation curve had the shape it did, and why it folded over at high frequencies. Quanta of high energy will be fewer because a large amount of energy is required to form them, and although quanta of low energy will be numerous, they won't add up to much because of their low energy. Most of the energy will be in between at intermediate wavelengths.

Planck was not entirely satisfied with his interpretation of his formula and was hesitant about promoting it. For years he continued to believe it was just a happy accident and would eventually be explained and brought into classical physics. But such was not to be. Indeed, he eventually received the Nobel Prize for his work, but it took until 1919. He was nominated for the award in both 1907 and 1908 and nearly got it in 1908, but there was still too much controversy at the time.

EINSTEIN TO THE RESCUE

For several years after Planck announced his formula, few took it seriously. One who did, however, was Albert Einstein. He was twenty-one years old at the time of its publication and had become interested in heat radiation while still a student. In his second year at the Zurich Polytechnic, he had studied Kirchhoff's book along with his regular curriculum. In his third year, his teacher, Heinrich Weber, presented the problem in class. Weber had actually made several measurements of the energy spectra of heated objects.

When Planck's paper appeared, Einstein was having a bad year. He had just graduated and was searching for a job. After a couple of discouraging years, he finally found one at the patent office in Bern, Switzerland, in 1902. Einstein had no doubt studied some of Planck's earlier papers; in fact, the paper containing his famous formula appeared in the same issue in which Einstein had his first published paper, a paper on *capillarity* (the rising of water or other fluid in a very fine tube).

Einstein's reaction to the paper was twofold. He was excited about it but had reservations about the way it had been derived. He was sure Planck had made an incorrect assumption; in addition, he had based his calculation on formulas which Einstein had reservations about.

Planck had assumed the radiation in the blackbody cavity was being emitted by *resonators*. He viewed these resonators as massless springs with an electric charge attached to them. Maxwell had shown that radiation was emitted when an electric charge underwent acceleration, and oscillation was a form of acceleration. The springs were all of different stiffness and could oscillate freely. When heat was applied to the cavity, the springs were set in motion and the accelerating charges emitted radiation according to their frequency. The oscillators eventually came to equilibrium in which they absorbed and emitted radiation by the same amount. Classically, the resonators could oscillate at any frequency and gave rise to a continuous spectrum of radiation. Planck assumed, however, that the resonators were quantized and gave out radiation only in discrete amounts. Of particular importance to him, it was the resonators that were quantized; the radiation was still electromagnetic waves and in theory could have any frequency. Their apparent discreteness as they were emitted from the cavity was due to the quantization of the resonators.

Einstein went a step further. To understand his daring move, we have to fill in a bit of background. He had been studying a phenomena that was discovered by Heinrich Hertz in 1888. At the time the significance of Hertz's observation was not fully appreciated. Only later, when Philipp Lenard of Germany looked at the effect, did its true significance come through. Lenard had been an assistant to Hertz. In 1902 he began studying what we now call the *photoelectric effect*. He showed that an electric effect was produced by light falling on certain metals, and it was caused by the emission of electrons from the surface of the metal. In particular, only certain frequencies (colors) of light could produce the emission of electrons. Also, for a particular frequency, electrons of fixed energy were given off. But of most importance, and strange at the time, was the fact that the energy of the electrons that were emitted did not depend on the intensity of the light. As the light was made brighter, the energy of the electrons stayed the same. Their number increased, but not their energy. Their energy appeared to depend only on frequency, and this didn't seem to make sense.

Lenard received the Nobel Prize in 1905 for his work. His early brilliance was, however, overshadowed in his later life. He made a fool of himself in his strong support of Hitler and his denouncement of the Jews. He tried, in vain, to show that Einstein's theory of relativity was wrong, even though he did not understand it. In addition, he tried to show that others had arrived at certain aspects of the theory before Einstein. All of this came, of course, many years after Einstein began to take an interest in the photoelectric effect.

Einstein was intrigued by Lenard's discovery and set out to explain it. In a letter to his fiancée, Mileva, he wrote, "I have just read a wonderful paper by Lenard on the generation of cathode rays by ultraviolet light. . . . I am filled with such happiness and joy that I must absolutely share some of it with you."[3]

Einstein found he could explain the photoelectric effect by taking Planck's work to heart, and extending it. Planck had assumed that the resonators were responsible for the discreteness of the radiation. Einstein said no. He suggested that the radiation itself was discrete; in other words, it was quanta. We now refer to this radiation as *photons*. Ordinary light is a good example of photons of a particular wavelength. This was obviously a giant and daring step. As I mentioned earlier, Maxwell had shown that light consisted of waves and was a continuous phenomenon, and his theory had been universally accepted. It was almost blasphemous to suggest otherwise in light of the large number of experiments that had verified Maxwell's theory.

According to Einstein, the photons of light penetrated the metal's surface and were absorbed by electrons within the metal. If, upon absorption of a photon, an electron had enough energy to escape, it would break free and fly out of the surface. A certain amount of work would be needed to free itself. We now refer to this as the *work function* of the metal. Each metal had a different work function. So when the electron was emitted it would have the energy of the photon minus the work function.[4]

This paper was the first of Einstein's five famous papers of 1905. Most now refer to 1905 as a "miracle year" in the history of physics. Two of his papers proved the existence of atoms and molecules, and one of them was the celebrated paper on relativity that gave us a new view of space and time. It is interesting that in writing to a friend, Einstein characterized only his paper on the photoelectric effect as revolutionary.

Einstein opened his paper by pointing out an inconsistency. He did this frequently in his early papers. He wrote, "A profound formal difference exists between current theories of matter, in which the energy of a body is represented in a discrete way, and Maxwell's theory, in which the energy is a continuous function of fields." He suggested that the discrepancy could be resolved by assuming that the radiant energy is distributed discontinuously. In other words, it is discrete and consists of a finite number of quanta. His statement, "Indeed, it seems to me that observations of 'Blackbody' radiation . . . and other related phenomena associated with the emission and transport of light appear more readily understood if one assumes that the energy of light is discontinuously distributed in space," is as revolutionary as Planck's earlier statement was.[5]

As might be expected, Einstein eases us into his surprising suggestion by praising Maxwell's theory of light. "The wave theory of light, which operates with continuous spatial functions, has proved itself superbly . . . and will probably never be replaced by another theory." Despite this, he goes on to suggest an alternative.

The paper was submitted to *Allenen der Physiks* in late 1905. Planck was the editor of the journal and the one who made the decision as to whether papers were acceptable. He accepted the paper without any comments. Papers are frequently sent back to authors for revision before they are published, but this one was not.

We do know, however, that Planck didn't think much of the paper, and he didn't take it seriously. As late as 1913, when he nominated Einstein for membership in the Prussian Academy of Science, he wrote, "that sometimes, as for instance in his hypothesis of light quanta, he may have gone overboard in his speculation [but it] should not be held against him too much."[6]

Scientists didn't know what to make of Einstein's suggestion, and few took it seriously, but it did make a prediction related to the photoelectric effect. Lenard's studies were not sufficient to verify Einstein's relation, and for another decade there was considerable doubt about it. But by 1914 a considerable body of evidence had accumulated that tended to support it, and finally in 1916 Robert Millikan of the United States verified the relation beyond any doubt.

Einstein's introduction of the particle concept for light created a serious problem. What was light? Was it a particle or a continuous wave? For many years this would be a problem. Scientists liked to joke that they used the wave theory on Monday, Wednesday, and Friday, and the particle theory on Tuesday, Thursday, and Saturday.

The problem didn't bother Einstein. He went on to use the idea to treat the *specific heats* of solids at varying temperatures and was able to explain certain facts about specific heat (for example, the specific heat of certain solids decreases as the temperature decreases).[7]

Einstein received the Nobel Prize for his introduction of the idea of light quanta and its explanation of the photoelectric effect. Many thought he should have received it for his theory of relativity, but it was still too controversial in 1921, the year he received the prize.

The Bohr Atom

Rutherford had given us the nuclear atom. Planck had introduced the quantum, and Einstein had made it more convincing. But many problems remained. One of the major ones was that no one knew where the electrons were in the atom and what role they played. According to Rutherford they were in a cloud around the nucleus, but other than that little was known. Indeed, many were still convinced that Thomson's model was the correct one. Everything was there, but someone was needed to bring it all together into a clear, sharply focused picture. That person was Niels Bohr.

Bohr was born in Copenhagen on October 7, 1885. He had an older sister, Jenny, and a brother, Harald, who was two years younger. His father was a professor of physiology at the University of Copenhagen. Bohr received a good education, both at the public schools and at home from his education-minded parents. He was encouraged to join in the conversation, which covered topics from science to politics.

Bohr was very close to his brother, who was regarded

by most as the brightest of the children. But Niels's father didn't agree; he frequently referred to Niels as the "thinker" of the family.[1]

Niels was an excellent student, but he was not considered a genius. He was relatively poor in composition, and it plagued him throughout his life. He had difficulty writing scientific papers and usually wrote numerous drafts before he was satisfied. Even then his early papers were not well organized, and most were difficult to read. Despite his difficulty with composition, he did well enough in other subjects, particularly mathematics and physics, to place first in his class.

Unlike many scholars, he did not spend all his time with his nose buried in books. He excelled in athletics, particularly soccer and skiing. And even in later life, he was almost unbeatable in Ping-Pong.

His first taste of research came when he was nineteen. The Royal Academy of Science of Copenhagen offered a gold medal for the best paper on a selected research topic. In 1905 the topic was to determine the *surface tension* (a force exerted across the surface of a liquid) of a number of liquids by measuring the waves produced when they were forced through a small opening. Bohr worked diligently on the project for over a year and won the gold medal. His paper was 114 pages long.[2]

Early on it appeared as if Harald was going to be the most successful member of the family. His soccer skills developed well beyond those of Niels, and he made the 1908 Olympic soccer team that won a silver medal at the games in London. He also completed his master's degree in mathematics several months before Niels. Niels completed his master's degree in the spring of 1909 with a thesis on the theory of metals. Upon completion he immediately began working on his doctorate, continuing with the research on the theory of metals he had begun earlier. He was now also taking an interest in atomic models, particularly Thomson's model.

His thesis defense was announced in the newspapers, and the hall where it was held was packed. It's unlikely that anyone in the hall, including most of his committee members, understood the details of his thesis, but he passed with flying colors and was awarded a one-year scholarship to study abroad. He decided to go to Cambridge to work under J. J. Thomson. Many of the best physicists in the world were at Cambridge, and Bohr eagerly anticipated his visit.

Niels Bohr

He arrived in September 1911. With all the fanfare he had got in Denmark when he graduated, he was somewhat disappointed by the reception he received in England. He was now just another visitor; furthermore, he was still having trouble with the English language, so it was difficult for him to communicate. He wrote Harald telling him that Thomson was polite at their first meeting, but very busy.

Bohr had translated his thesis into English and had given it to Thomson for his comments. He was worried because he had criticized Thomson's atomic model in his thesis and had pointed out several errors. Thomson was so busy that it's unlikely he ever read it in detail. He did notice the criticisms of his atomic model, however, and talked to Bohr about them briefly.

Bohr was disappointed and kept hoping that Thomson would sit down with him and discuss his thesis work. But all Thomson did was suggest that he try to get it published. Unfortunately, when he tried, the publisher told him it was too long; he would have to cut it down to half the length before they would consider it. As a result, it was never published.

Bohr's biggest disappointment at Cambridge was that Thomson didn't take him seriously as a theorist. He gave him an experimental project dealing with positive rays (canal rays). Bohr's heart wasn't in the work, and he made little progress. In the meantime, Rutherford, who was then at the University of Manchester, stopped in. He was on his way back from a conference in Brussels. He gave a talk at Cambridge, and there was a dinner in his honor. Bohr didn't manage to talk to him, but he was impressed, particularly impressed with Rutherford's enthusiasm.[3]

Bohr began thinking about going to Manchester to work with Rutherford, but he needed an introduction. He remembered that a friend of the family lived in Manchester and was sure that he would know Rutherford. Bohr visited him and got his introduction. Bohr talked to Rutherford for several hours, telling him of his desire to come to Manchester. Rutherford was impressed with Bohr and invited him to come, but he told him not to be too hasty about leaving Cambridge.

It is, perhaps, surprising that Rutherford took a liking to Bohr. Rutherford had a well-known disdain for theorists, and for the most part did not take them seriously (even Einstein's work didn't impress him). But he was an avid soccer fan and was delighted when he found out that Bohr had played soccer. He was doubly delighted to hear that Harald Bohr, who was now famous for his exploits in the Olympics, was his brother. "Bohr's different," he would say when asked about his interest in a theoretician. "He's a soccer player."

Bohr returned to Cambridge after the meeting, still uncertain. He thought about Rutherford's advice about leaving Cambridge, but he still wanted to go to Manchester. He remembered that his brother was coming for a visit the following week; he decided to ask him for his opinion. Harald encouraged him to go, and in March 1912 Bohr arrived in Manchester. From the beginning he was much happier, and he found the scientific climate much more to his liking. Many of the leaders in the field of radioactivity were at Manchester: Hans Geiger, Ernest Marsden, and George de Hevesy. Hevesy eventually became one of his best friends. Bohr found Rutherford quite different from Thomson. Rutherford encouraged him, took an interest in his work, and spent hours discussing his ideas with him. In a letter to Harald, he wrote, "Rutherford is a man you can rely on and he comes regularly to inquire how things are going, and he talks about the smallest details."[4]

Rutherford encouraged him to work on an experimental project, but Bohr soon convinced him he would do much better on a theoretical project. Indeed, Bohr had told Rutherford of his tremendous enthusiasm for his nuclear model of the atom, and Rutherford was no doubt flattered. He had received considerable criticism for it and was no doubt anxious to prove its worth.

Within a couple of months it was obvious that Rutherford had considerable confidence in Bohr's ability and knowledge. On one occasion he was standing with Bohr when someone came up and asked him a question. Not sure of the answer, he shrugged and said, "Ask Bohr." The person turned and asked Bohr the question, and Bohr answered it without difficulty.

Bohr's first project was a calculation of the energy losses of alpha particles as they passed through matter. Alpha particles had always been a favorite of Rutherford's, and he was impressed with what Bohr was able to do. Of particular importance for Bohr, however, his work with alpha particles gave him several ideas about how to proceed with his theoretical model of the atom. By July he had decided how he was going to deal with the electrons. Work by Richard Whiddington on cathode rays had shown that radiation was absorbed and emitted only at certain energies. To Bohr this meant that the electrons had to be in specific orbits, and each had to have a particular energy. In short, it was like a miniature solar system.

But his major problem was still the instability of the model, and it was the reason so few took Rutherford's model seriously. Maxwell had shown rather decisively that an accelerating electron would give off radiation, and as a result it would lose energy. And there was no doubt about the electron in Rutherford's model; if it was in orbit around the nucleus, it had to be accelerating, and if it was accelerating, it had to be giving off radiation. In fact, it would be giving off radiation continuously. Futhermore, it was easy to show that it would lose energy at such a rate that within a tiny fraction of a second it would spiral into the nucleus.

It was obvious that this didn't happen. Atoms were stable. Bohr realized that something drastic was needed, and this drastic action would likely involve the *quanta* that Planck had introduced.

Classical physics obviously could not be obeyed if Rutherford's model was to work, but Planck and Einstein had both shown that classical physics isn't always obeyed. In Bohr's case, however, there were several challenges. First, he had to get rid of the continuous radiation. Atoms didn't radiate continuously. Unless they were influenced from the outside, they didn't radiate at all; furthermore, when they did radiate, they gave off energy discontinuously. This was evident in the spectral lines that arose when they were heated to incandescence. These lines corresponded to different frequencies, and they appeared to be irregularly spaced.

With his idea of orbits and energy levels, Bohr was sure he was on the right track, but his scholarship was rapidly running out. It was May and he would have to return to Denmark at the end of July. He shifted into high gear during June and July, frequently working late into the night.

He decided to concentrate initially on the simplest atom, hydrogen. It would have a simple nucleus, consisting of a proton, and around it would orbit a single electron. The charges of the proton and electron were equal and opposite, so the electron would be held in orbit by its motion and the attractive force of the proton. Although there was only a single electron, Bohr decided there had to be several orbits, or energy levels, for it to reside in (fig. 3). This was in line with the discreteness of the quantum postulated by Planck. Higher levels would have greater energy. Although it was contrary to the ideas of Maxwell, he postulated that the electron gave off no radiation when it whirled around the nucleus in its orbit; it gave off (or absorbed) radiation only when it jumped between orbits. In particular, when it jumped down from a higher orbit, or energy level, to a lower one, it gave off radiation with a frequency that corresponded to the difference in energy of the levels. Energy was related to frequency through the relation that Planck had given.

To many this was outrageous. It was nonsense. Bohr knew what the reaction would be, and he knew he would have to prove his assertions. His time at Manchester was now up, and he had to return to Copenhagen. He was married in August and came back

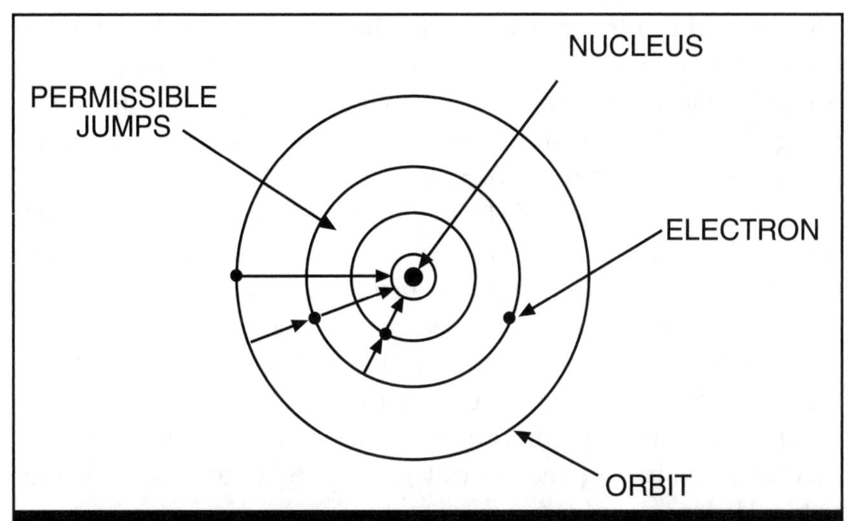

Fig. 3. The Bohr atom showing the various orbits around the nucleus

to England briefly on his honeymoon, but little was accomplished on his model. His major aim now was to prove his ideas.

After his honeymoon he took a teaching post at the University of Copenhagen. It was a junior position, and he was given a heavy teaching load. No professorship was available; he would have to wait until someone retired for that. A professorship would have allowed him considerable time for research, but as it was he was loaded down teaching medical students, which frustrated him. He had very little time left over for his theoretical work.

In January 1913 he was continuing his search for proof of his ideas when an old friend from his student days visited him. His name was Hans Hansen, and he was an experimentalist in spectroscopy. Bohr talked to him about his work. At one point Hansen asked, "Does your model explain the Balmer series of hydrogen and the Rydberg constant?" Bohr wasn't sure what he was taking about. He had taken some spectroscopy in his undergraduate courses, but he didn't have a detailed knowledge of what was going on in the area.[5]

The science of spectroscopy was well developed at the time. Thousands of lines from numerous elements and compounds had been observed and measured. Things were so complex, though, that Bohr had decided that spectroscopy would be of little use to him. He had to admit to Hansen that he hadn't heard of the Balmer series and hadn't thought of using his theory to explain it. He immediately went to the library, however, and looked up everything he could on Johann Balmer, Johannes Rydberg, Walter Ritz, and others who had tried to explain spectra.

"The moment I saw Balmer's formula, I knew it was what I needed," he said. Within a short time he was able to write down an expression for the Balmer series using his model. Everything fit together perfectly. Indeed, he was able to calculate a value for a constant called the *Rydberg constant* that had been measured experimentally. Bohr derived an expression for it in terms of fundamental constants of nature, and he hit the number on the nose. His theoretical value agreed almost exactly with the experimental value. He was delighted.

He started writing a paper for publication, but soon discovered it would be far too long for a single paper, so he divided it into three parts. The paper, which he titled "On the Constitution of Atoms and Molecules," would eventually become a classic. He sent the first draft of part one to Rutherford in March 1913, asking for his advice and recommendations, and if he were satisfied with it to pass it on to *Philosophical Magazine* for publication.

Rutherford was impressed, but he found it difficult to accept some of Bohr's strange ideas. Furthermore, he thought the paper was much too long, and he wrote Bohr suggesting that he shorten it. Bohr had spent countless hours sweating over many drafts, agonizing over each word, and he was very reluctant to change anything. He wrote Rutherford of his reluctance and the importance of keeping every word. Rutherford finally agreed and passed it on to *Philosophical Magazine* on April 5. It appeared in the magazine in July and was concerned mainly with hydrogen and its line spectra.

The second and third installments of the article came in June and August. The second paper dealt with systems having a single nucleus but more than one electron. It was published in September. The third section dealt with molecules, and it appeared in November. In this paper Bohr attempted to account for the periodic table of elements, and he discussed x rays and radioactivity.

Bohr had an ace in the hole—his calculation of Rydberg's constant—but he needed more. He still had to relate his ideas to classical theory. After all, classical theory had been used for years and was known to be exceedingly accurate. Looking at the structure of the energy levels he had obtained for the hydrogen atom, he saw that as they got farther and farther from the nucleus they also got closer together. Indeed, the outer levels were so close together they looked like a "continuum" of energy states. In short, the changes in energy were virtually continuous. This was, of course, the situation in classical theory according to Maxwellian electrodynamics. Radiation was given off in a continuous mode and could have any value.

Bohr calculated what the energy output would be for the outer levels according to Maxwell's theory and compared them with what he got from his theory. The agreement was amazing and was exactly what he needed. This is now referred to as the *correspondence principle*. It states that when quantum states become very dense (i.e., close to continuous) there is exact agreement with classical theory.

REACTION TO BOHR'S THEORY

Bohr anxiously awaited a reaction to his theory, and initially most of it was negative. Many well-known scientists expressed skepticism, but there were bright spots. Bohr had an opportunity to present his theory at the yearly meeting of the British Association for the Advancement of Science, which was held in Birmingham in September. His papers had already attracted considerable attention, so there was some anticipation. James Jeans, who was a strong supporter of the theory, gave the lead presentation in which he surveyed many of the problems of radiation and electrons. He said, "Dr. Bohr has arrived at a most ingenious and suggestive, and I must add convincing, explanation of the laws of spectral series."[6]

Lorentz asked Bohr about the connection between his postulate and classical theory. Bohr merely replied that the theory was not yet complete, so the answer was not yet known. Lorentz didn't appear to be impressed. Neither was Lord Rayleigh, who said, half jokingly, " A man over sixty should not take part in a debate over new problems." Thomson's view was also less than encouraging, which can perhaps be understood. He was still clinging to his own model, and he said that he did not believe in a "quantum" model of the atom. He felt his own more traditional model explained things better.

Others also protested later. Max Von Laue of Zurich said, "This is nonsense. Maxwell's equations are valid under all circumstances." And Max Born of the University of Göttingen expressed considerable doubt as to its validity.

Bohr did receive considerable support from George de Hevesy of the University of Manchester, whom he had worked with earlier. Hevesy wrote a paper dealing with his research on radioactivity and reported that his results agreed with Bohr's theory.

Of course, everyone was interested in Einstein's opinion. He was not at the Birmingham meeting and heard about the theory at the Assembly of German Natural Scientists. Hevesy was at the meeting and reported to Bohr. Einstein told him that the theory was very interesting and important if it was right. Hevesy told Einstein that the theory predicted the Pickering-Fowler spectrum of helium. Einstein looked surprised. "Then the theory . . . must be right," he said.[7]

In regard to the Pickering-Fowler spectrum of helium, the English spectroscopist Alfred Fowler had pointed out a small discrepancy to Bohr between the experimental results he had got and Bohr's theory. Bohr looked into the problem and made a new calculation taking into account the motion of the nucleus around the center of gravity in helium. With this correction there was good agreement with Fowler's results.

The first experimental support of Bohr's theory came from the work of H. G. Moseley. In the summer of 1913 Moseley began investigating the emission of x rays from various substances. He was soon able to determine the wavelength of the emitted x rays, and within months he was able to show that his results agreed with those predicted by Bohr's theory. Moseley's work gave the first experimental proof of Rutherford's nuclear atom, and in the process it confirmed Bohr's positioning of the electrons.

Another important verification came from the work of the German physicist Johannes Stark. Stark showed that a spectral line is split into several lines when the emitting atom is placed in an electric field. Pieter Zeeman had shown the same phenomenon a few years earlier for the case of a magnetic field. Rutherford drew Bohr's attention to Stark's paper, which appeared in November 1913, and Bohr went to work immediately to show that the effect could be explained by his theory. Within weeks he had succeeded. His results were published in the March issue of *Philosophical Magazine*.

In 1914 two German scientists, James Frank and Gustav Hertz (a nephew of Heinrich Hertz), found that mercury atoms bombarded with electrons absorb energy from the electrons. But the energy was not absorbed below a certain critical value; above this value, however, it was easily absorbed. Frank and Hertz made no mention of Bohr's theory in their paper and gave their own interpretation of their results. They assumed they had *ionized* (given them a charge) the outermost electrons in the mercury. When Bohr heard of the paper, he realized immediately that their interpretation was wrong. He suggested a new interpretation based on his theory: the energy change was not a result of ionization, but rather the absorption of quanta. It's interesting that Frank and Hertz continued to cling to their explanation for another two years. Finally, however, they realized Bohr was right.

Even though Bohr was now becoming well known and respected for his new theory, a professorship at the University of Copenhagen was still not forthcoming. He continued with his heavy teaching load. He informed Rutherford of his difficulties, and as it turned out George Darwin's readership position at Manchester had expired and he was leaving. Rutherford offered the position to Bohr and he accepted.

Before Bohr left, he and Harald went on a walking tour of the Alps and Germany, where they visited Göttingen, Munich, and Würzburg. During his visit Bohr was able to present his theory to many well-known physicists. As their tour came to an end in August, World War I broke out, and Niels and Harald had to rush to get out of Germany before the borders were closed.

Bohr arrived in England in October to take up his position at Manchester. He remained there until 1916. During this time he considered the possibility of elliptical orbits in hydrogen, which enabled him to explain a splitting, or doubling, of hydrogen lines that had been observed.

One of the strongest supporters of Bohr's theory was the German physicist Arnold Sommerfeld. Sommerfeld also dealt with elliptical orbits and applied Einstein's special theory of relativity to the hydrogen atom.[8] He presented his modification of Bohr's theory to the Munich Academy of Science in December 1915 and sent a copy of his papers to Bohr. Bohr received them enthusiastically. "I thank you so much for your most interesting and beautiful papers. I do not think that I ever have enjoyed the reading of anything more than I enjoyed the study of them."[9]

As the war continued, Bohr's fame continued to spread. He received an invitation to lecture at the University of California, but had to decline because of the war. In Denmark many were beginning to worry that they might lose Bohr to England permanently, and a professorship at the University of Copenhagen was finally offered to him in March of 1916.

In autumn 1916 Bohr returned to Copenhagen to take up his duties as professor. Within months he had an assistant by the name of Hendrik Kramers. Kramers proved to be exceptionally able and remained with Bohr for ten years. Bohr's theory had been successful in the case of the hydrogen atom, but the extension to helium posed serious problems. Together with Kramers, Bohr now attacked the problem of the helium atom. They soon saw, however, that serious modifications would be needed.

One of the major successes of Bohr's theory was that it put chemistry on a firm footing. Chemistry is concerned with how atoms react and how they combine to form molecules. Bohr explained this using his energy levels. He was also able to explain the periodic table of elements. He showed that only two electrons could go into the first energy level (in chemistry, levels are usually referred to as *shells*). The third electron would therefore have to go into the second shell. Bohr also showed that only eight electrons could go into the second shell, and it was the outer (or *valency*) shell that determined the chemical properties of the elements. If there was a lack of electrons in this shell, the atom could "share" electrons with another atom. In this way, two elements could join to form a molecule. Indeed, all chemical reactions can now be explained as a "sharing or swapping" of electrons between atoms.

THE CRAZY IDEA: WAVES OF MATTER

Bohr struggled with his theory for years, but it eventually became obvious that something else was needed, and when it came, it came from a most unlikely source: a French prince, Louis de Broglie. De Broglie was born into a noble French family on August 15, 1892. He was educated at the Sorbonne, obtaining a degree in history in 1913. His brother Maurice was a well-known experimental physicist who eventually interested him in physics. De Broglie became particularly intrigued with Einstein's theories. Before he could do anything about it, however, World War I broke out and he was drafted. He spent most of the war stationed in the Eiffel Tower as a radioman.

Louis de Broglie

When the war ended, de Broglie decided to pursue his interest in physics, and he went to the University of Paris in pursuit of a doctorate. One of the major problems at the time, as we saw earlier, was the dual nature of light: it could be considered both a particle and a wave. De Broglie soon became fascinated with waves, particularly after he read several papers on the phenomenon by Marcel Brillioun of the College of France. Brillioun had studied the periodic motion of vibrating particles in an elastic medium. He even went as far as trying to relate his work to Bohr's atom, but was not successful.

De Broglie wondered if it was possible that the electron, in addition to its particle properties, also had wave properties. After all, light was both a wave and a particle. Furthermore, if it did, could it account for Bohr's orbits? At the end of summer 1923, his ideas finally began to gel. To see what he concluded, consider a string. If we tie it down at both ends and jiggle it appropriately, we can set up a "standing wave" on it as seen in figure 4. The simplest wave of this type will have one loop. This is one half of the *wavelength* of the wave (the wavelength is defined as the distance from one point on the wave to the point where the wave is repeated exactly at a given time). The end points are referred to as *nodes*.

If we jiggle the string differently, we can create three nodes and two loops (fig. 5). In the same way we can continue to four nodes and three loops. The sequence can be extended indefinitely.

Was it possible that the electron also oscillated in the same way? In other words, could it be considered to be a wave with one, two, three, or more loops? If so, Planck's quantum *h* would somehow have to be involved. De Broglie decided to equate the mo-

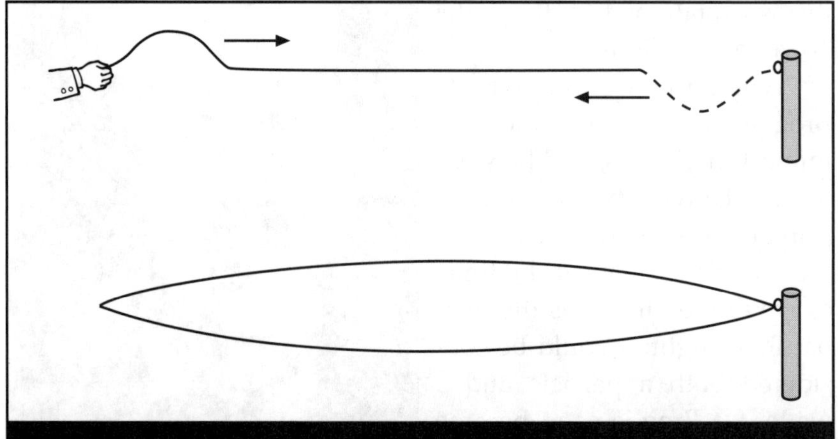

Fig.4. The top illustration shows a moving wave in which the hump moves back and forth on the string. The bottom one shows a standing wave of one-half wavelength.

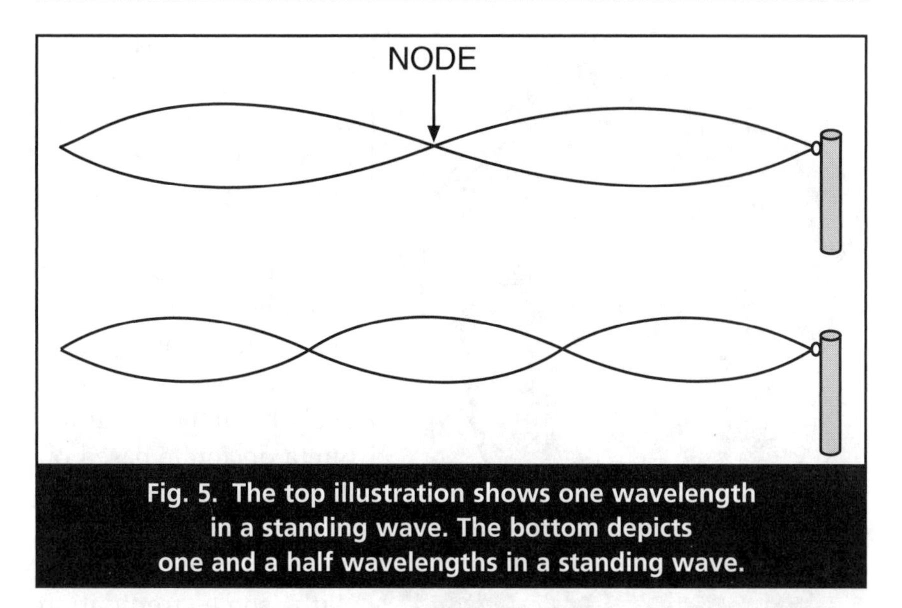

Fig. 5. The top illustration shows one wavelength in a standing wave. The bottom depicts one and a half wavelengths in a standing wave.

mentum of the electron to *h*, or more exactly, to *h* divided by the wavelength of the electron wave. He then compared these lengths to the lengths of Bohr's orbits and was pleased to find they were the same. The wave would have to be looped around in a circle, but the idea worked (fig. 6). He had explained the reason for the discrete sizes of Bohr's orbits: the electrons, which were standing waves, could fit only if they were a certain length. An integral number (i.e., 1, 2, 3 . . .) of standing waves had to fit into the orbit for it to be stable.

De Broglie published his results, then used them as part of his doctoral thesis. He presented his thesis to the faculty of the University of Paris on November 29, 1923. His thesis advisor was Paul Langevin. The examining committee praised the originality of the thesis and was impressed with the mathematics, but was uncertain about the proposal he had made. It was difficult for them to accept the idea of matter waves. One of the members of the examining committee, Jean Perrin, asked if the waves could be verified experimentally. De Broglie replied that they should be visible in diffraction experiments of electrons by crystals. These are experiments in which high-energy electrons are projected at perfect crystals and the angle at which they are scattered is measured.

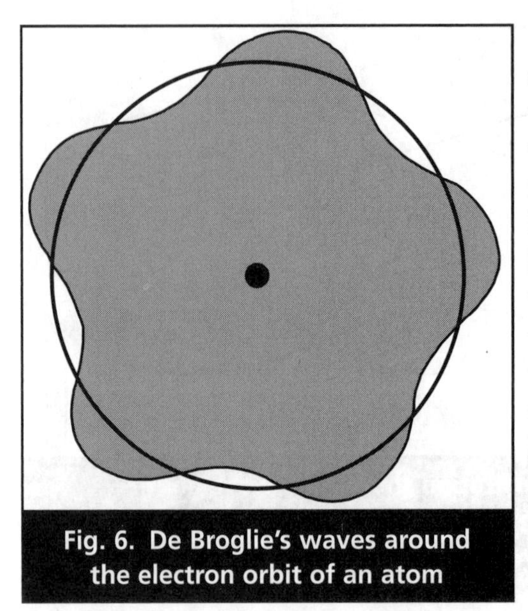

Fig. 6. De Broglie's waves around the electron orbit of an atom

The committee was in a quandary. De Broglie was from a noble family, with relatives who had made important contributions to France—his brother, Maurice, was a well-known experimentalist. They had to take the "crazy idea" seriously, but if they awarded him a doctorate based on it, and later it turned out to be incorrect, it would be an embarrassment. With some trepidation they decided to accept it.

In April 1924 de Broglie's supervisor, Langevin, talked about de Broglie's work at the Solvay Congress in Brussels, and he informed Einstein about it. Einstein was interested and requested to see de Broglie's thesis. In December 1924 he received a copy. Einstein was impressed. "It may look crazy, but it is sound," he said.[10]

Einstein passed de Broglie's thesis on to Max Born at Göttingen, who discussed it with James Frank, the department head. Walter Elsasser, Born's student, also sat in on the discussion. Elsasser suggested a test using the scattering of electrons from metals, but Frank pointed out that it had already been done by Clinton Davisson in America. Indeed, Davisson and a colleague, C. H. Kunsman, had been scattering electrons from metals since 1919 at Bell Laboratories.

Elsasser wrote a brief paper on de Broglie's results and how they could be tested. Davisson read the paper but didn't think much of it. He already had an explanation of the scattering. (I should mention at this point that Davisson had not been using single crystals.) In April 1925 Davisson had an accident in his lab. A liquid-air bottle exploded while the nickel target was extremely hot. It splashed over the nickel and caused it to recrystallize. As a result it became a single crystal.[11] When Davisson resumed his experiment, he noticed that the scattering had changed. To his surprise, when he compared the new results with de Broglie's prediction, there was excellent agreement.

He then began a series of experiments that thoroughly verified de Broglie. In the meantime George Thomson, son of J. J., had begun a series of experiments in which he projected electrons at thin metallic films. He also verified de Broglie's formula. In 1937 Davisson and Thomson received the Nobel Prize for their work. It is amusing that J. J. Thomson, the father, was awarded the Nobel Prize for showing the electron was a particle and George Thomson, the son, was awarded the prize for showing it was a wave.

Heisenberg's Arrays

Bohr's theory of the hydrogen atom was well established. Sommerfeld had made important contributions to the theory, and it was now known as the Bohr-Sommerfeld theory. But as experimental techniques improved, it became obvious that the theory was limited. The agreement between theory and experiment for the hydrogen atom was excellent, but there were difficulties with the helium atom, and anything more complex than it seemed hopeless. Besides this, many scientists were disturbed by the theory. It seemed to be an incoherent mishmash of classical and quantum ideas. By the early 1920s many well-known scientists were expressing their discontent. In 1924 Max Born of Göttingen wrote, "At the most we possess only a few unclear hints." In 1925 Wolfgang Pauli wrote, "Physics at the moment is again very muddled." No one was sure what the problem was, but most scientists were convinced that a completely new theory was needed. And, indeed, within a short time an important breakthrough was made by Werner Heisenberg of Göttingen.[1]

HEISENBERG

Werner Heisenberg was born in Würzburg, Germany, on December 5, 1901. His father was a professor of Greek studies at the University of Wurzburg, and he had a brother who was a year and a half older than he. Werner's early life was pleasant, and a love of academics was encouraged by his father. Through much of his early life, there was serious competition between his brother and him, much of it in the form of mathematical games. Werner discovered early on that he was particularly good at mathematical games and could easily beat his older brother. This led to considerable friction between the two.

Partially because of these games, Werner developed a serious interest in mathematics early in his life. By the age of twelve or thirteen, he was already studying calculus on his own, and soon thereafter he was studying Einstein's theory of relativity. The mathematics in Einstein's books gave him little trouble. He also read the notoriously difficult book *Space-Time-Matter* by Herman Weyl. Despite reading many books on physics, his major interest was mathematics. He was particularly interested in the theory of numbers and spent several months trying to solve Pierre de Fermat's famous last theorem.

Werner Heisenberg

But his life was not all studies. He was an excellent athlete—a good skier and an excellent long-distance runner. One of his favorite activities was hiking. Throughout his teen years he was involved in several different youth groups and loved to stay out overnight hiking in the mountains. He was also an excellent pianist. By the time he was a teenager, he was playing Mozart and Schumann concertos.

When he was nine, his family moved to Munich where his father took a position at the University of Munich. His family was fairly well-off, but life was not always easy for Heisenberg. He was thirteen when World War I broke out, and because of a severe shortage of food, he had to leave school to work on a farm for several months during the latter part of the war. Despite the distractions and difficulties, he was always at the top of his class. He was particularly good at mathematics but also did well in other subjects. He graduated from the gymnasium (high school) in 1920. With his father being a professor at the University of Munich, it was natural for him to attend the same school.

COLLEGE YEARS

Heisenberg was interested in both mathematics and physics but had decided early on that he would major in mathematics at the university. At that time there were two options. A student could take the regular schedule of classes, or he could be assigned to a professor, in which case he would start research relatively soon. Students were not usually assigned to a professor until their third or fourth year, but occasionally outstanding freshmen were selected for the honor. Heisenberg had done extremely well in the gymnasium and was encouraged by his father to get an assignment with a professor. He suggested the mathematics professor Ferdinand von Lindemann, and arranged a meeting with him.

Lindemann was close to retirement age and not accustomed to accommodating freshmen. Rather reluctantly he agreed to meet with Heisenberg as a favor to the young man's father. The meeting went poorly from the beginning. Lindemann had a small dog in his office that started barking the moment Heisenberg came in, and it didn't stop. On top of this, Lindemann was hard of hearing, and Heisenberg had to repeat everything he said. Finally Lindemann asked him what books he had read in preparation for college. "*Space-Time-Matter*, by Herman Weyl," Heisenberg said, hoping it would impress Lindemann. But it had the opposite effect. Lindemann frowned, "In that case you are completely lost to mathematics," he said.[2]

Heisenberg left the meeting depressed. He wasn't sure what he wanted to do, but he now began thinking more seriously about physics. He talked it over with his father, and a meeting was arranged with Arnold Sommerfeld. Sommerfeld was well known and had made important contributions to theoretical physics. Furthermore he was director of the prestigious Institute of Theoretical Physics. Heisenberg visited Sommerfeld, and this meeting went well. As Lindemann did, Sommerfeld asked what books he had read. Heisenberg was a little more hesitant this time, but he answered, "*Space-Time-Matter*." Knowing how difficult the book was, Sommerfeld was impressed, and he agreed to let him into his seminar. Many of the best physics students in Europe were in this group.[3]

HALF-INTEGRAL QUANTUM NUMBERS

Heisenberg was assigned a schedule of classes, and within days he was called into Sommerfeld's office. Sommerfeld had been working on a problem related to the Zeeman effect, the splitting of spectral lines in a magnetic field. This splitting had not been explained and was not accurately predicted by Bohr's theory. Sommerfeld assigned him the problem of determining a rule for the splitting, based on quantum numbers. Two weeks later Heisenberg had the problem solved, and he took the solution to Sommerfeld. Sommerfeld was amazed, but when he looked through the solution, he was confused and a little disappointed. Heisenberg had indeed solved the problem, but in the process he had used half-integral quantum numbers. According to Bohr's theory, quantum numbers were integers; they could be 1, 2, 3, 4, 5 . . . but not a fraction of these numbers.

Another of Sommerfeld's students was Wolfgang Pauli. He was a senior in his fifth semester at the time. Later he would make important contributions to quantum theory. Sommerfeld showed him the theory and asked his opinion. Pauli shook his head, "Where will it all end?" he asked. If half-integral quantum numbers were allowed, Pauli was sure that other fractions would soon follow, and the whole scheme would be destroyed. Sommerfeld agreed with him and did not encourage Heisenberg to publish his results.

Wolfgang Pauli

Pauli eventually became a lifelong friend of Heisenberg's and was instrumental in steering him into atomic physics. Heisenberg originally wanted to do research in relativity theory, but Pauli had just written a book on relativity theory and cautioned Heisenberg that the prospects in the area were limited. Heisenberg therefore decided to concentrate on atomic physics.

Several weeks after Heisenberg brought his results to him, Sommerfeld got a letter from Alfred Landé of Frankfurt. Landé had enclosed a preprint of a paper he was planning to publish, and he wanted Sommerfeld's opinion on it. Landé was also working on the Zeeman effect, and to Sommerfeld's surprise he also used half-integral quantum numbers in his theory. In fact, his ideas were almost exactly the same as Heisenberg's. Landé was a well-established physicist, and Sommerfeld was reluctant to take a strong stand against his idea. He wrote back telling him that his student had come up with the same idea. The following month Landé published his paper. Heisenberg was depressed; he had been scooped. Sommerfeld consoled him, assuring him that the problem was far from solved. Landé had merely published a rule for determining the splitting. He encouraged Heisenberg to keep working on the theory.[4]

Heisenberg took his advice and within weeks he had a complete theory of the effect. He referred to it as his *core model*. But again Heisenberg had done something strange. He considered the nucleus and the inner electron shells to be a unit surrounded by the outer valency shell. The inner shells plus the nucleus were considered to be the *core* of the atom. Again using half-integral quantum numbers he was able to explain the Zeeman effect. And again Sommerfeld was hesitant. This time, however, he encouraged Heisenberg to publish his result. It was Heisenberg's first scientific paper. Sommerfeld sent a reprint of it to Niels Bohr in Copenhagen.

Bohr did not like the model. He was strongly against the use of half-integral quantum numbers, and he didn't like the idea of an atomic *core*. Pauli also stated that he thought the model was "repugnant."

VISIT TO GÖTTINGEN

After World War I, Bohr had managed to convince the University of Copenhagen to build an Institute of Theoretical Physics. It was a three-story building with a staff of eight. As director, Bohr's living quarters were on the third floor (he later moved to a house beside the institute). Over the next few years, almost all of the major physicists in Europe and many from America visited or worked at this institute, and it would play a major role in the development of physics.

Officials at the University of Göttingen invited Bohr for a series of lectures in June 1922, and since Göttingen was not far from Munich, Sommerfeld took several of his students to hear the talks. Among them was Heisenberg. Heisenberg was impressed with Bohr. On the last day of the lectures, Bohr talked about some work his assistant Hendrik Kramers had done. At the end of the lecture, he summarized the results and asked for questions.

Heisenberg had worked in the area, was familiar with what Kramers had done, and disagreed with it. When Bohr asked for questions, Heisenberg stood up and pointed out a serious difficulty in the theory. Bohr was taken back at first and wasn't sure what to say, but when he thought about it, he knew that Heisenberg was right. He was impressed with the insight that such a young student appeared to have—Heisenberg was only eighteen at the time. Bohr had heard of Heisenberg's core model but had never met him personally. He asked Heisenberg to meet him after the lecture.[5]

When Heisenberg went up to him after the lecture, Bohr suggested that they go for a walk. Bohr asked many questions as they walked and was soon thoroughly impressed with Heisenberg. They talked about the problems of the Bohr model of the atom and the Zeeman effect. Bohr was so impressed with Heisenberg's insight into the problems that he invited him to come to Copenhagen to his Institute of Theoretical Physics. Heisenberg was overjoyed. Bohr was, after all, aside from Einstein, the most famous physicist in the world, and at that time he was only thirty-seven years old. Arrangements were made for Heisenberg to visit at a later date.

MAX BORN

Max Born

Sommerfeld was invited to the United States for a series of lectures during the winter of 1922–23. He got in touch with Max Born of Göttingen and asked if Heisenberg could visit for a semester while he was gone. Born agreed. Sommerfeld was sure the experience would be good for Heisenberg because Göttingen had many excellent teachers and was a leading institute in both mathematics and physics. Worried that he might lose his star pupil, Sommerfeld made Born promise to send him back at the end of the semester.

Within a few weeks, Born was as impressed with Heisenberg as Sommerfeld was. "I have become very fond of Heisenberg," Born wrote to Sommerfeld in the United States. "He is very well liked and highly regarded by all of us. His talent is unbelievable, but his nice, shy nature, his good temper, his eagerness, and his enthusiasm are especially pleasing." While he was at Göttingen, Heisenberg continued to work on his core model.[6]

Heisenberg was overwhelmed by Göttingen at first, but he soon grew homesick. In Munich he was still living with his parents, and this was the first time he had ever been on his own for any length of time. He was also separated from many of his friends. He was therefore relieved when Sommerfeld got back, and he anxiously returned to Munich. He was now in his sixth semester and ready to take his exams for a doctorate.

THE DOCTORATE EXAM

It's amazing that Heisenberg had spent only three years at the University of Munich, yet he had published several important papers and was ready to defend his doctoral thesis in July 1923. Interestingly, his doctoral thesis was not on atomic physics. Sommerfeld had two interests: atomic spectra and *hydrodynamics* (the study of the dynamics or motions of fluids). Heisenberg was attracted to hydrodynamics because of the complex mathematics of the theory and the interesting predictions that could be made.

Sommerfeld assigned Heisenberg a problem related to the transition from laminar, or smooth, flow to turbulence, in fluids. Osborne Reynolds of England had done considerable work in the area, but Sommerfeld was sure that a more satisfactory solution would be found. Heisenberg did, indeed, make several important advances. He felt confident as he went into his exam, but little did he know it would be a disaster for him. His committee consisted of Oskar Perron of the mathematics department, Hugo von Seeliger of the astronomy department, and Sommerfeld and Wilhelm Wien jointly representing the physics department. Wien was a well-known experimental physicist. As it turned out, Sommerfeld and Wien were not the best of friends, and had had a long feud. Sommerfeld regarded experimental physics as secondary and inferior to theoretical physics, much to Wien's consternation. Furthermore, Wien had insisted that Sommerfeld's students have some exposure to experimental physics. To console him, Sommerfeld had his students take Wien's class in experimental physics.

Heisenberg had taken the class and had not enjoyed it. Wien had assigned him an experiment to do related to the Zeeman effect, but he had given him no instructions on how to do it. One of the instruments used in the experiment was the Fabry-Perot interferometer. The lab was reasonably well equipped, but Heisenberg was not familiar with where things were kept and he never inquired. After considerable difficulty, he finally managed to complete the experiment. He did the best he could but knew it was less than satisfactory. In particular, he had not understood everything he did.

Despite the difficulties in his experimental physics class, Heisenberg was not worried. After all, it was well known that most theoretical physicists were not good experimentalists. He knew he had done an excellent job on his thesis, and it seemed unlikely that any of the questions in the exam would stump him. He was sure he was well prepared. As expected, he had no problems with Perron's math questions or with Seeliger's astronomy questions. But then Wien began asking him about the experimental work he had done with the Fabry-Perot interferometer. He began by asking him to derive an expression for the resolving power of the Fabry-Perot interferometer.[7] This was something Heisenberg had not bothered with and knew little about. He struggled for a while and finally had to admit he couldn't do it. Annoyed, Wien then asked him to derive a similar expression for the microscope. This was, of course, unfair, since it was obvious he knew little about the area. Heisenberg couldn't do it, but Wien persisted. He then asked him to derive the same expression for the telescope.

Again, Heisenberg admitted he couldn't. Wien was, by now, thoroughly annoyed. As a final question, Wien asked him how the storage battery worked. Heisenberg wasn't sure.[8]

Wien was so outraged he threatened to fail him. Sommerfeld, who considered Heisenberg to be a mathematical genius, was equally outraged, and embarrassed. After some discussion Wien agreed to give him the lowest passing grade (something like our D); averaged with Sommerfeld's A, his overall grade was C (equivalent to our C) and he passed. Heisenberg was thoroughly embarrassed and dejected to think he got such a low grade. Despite his low grade in the physics section of the exam, he was still graduated cum laude because of his excellent thesis.

That evening Sommerfeld had a small party to celebrate Heisenberg's graduation, but Heisenberg was not in the mood for celebrating. He excused himself early and left. Born had offered him a position at Göttingen when he graduated. Heisenberg therefore took the train the next day to Göttingen. He told Born of the debacle, then rather sheepishly asked, "Do you still want me?" Born had already made up his mind about Heisenberg's talents and wasn't worried. He asked about the questions that Wien had asked. Not being an experimentalist himself, Born sympathized with Heisenberg. When Heisenberg finished telling him about the questions, he said, "They were tricky." He told Heisenberg that he still wanted him and to report for work in the fall. Heisenberg was overjoyed. Over the summer he went on an extended hiking trip through Finland with several of his friends, and early in October he was back in Göttingen.[9]

TO COPENHAGEN

Heisenberg got along well with Born, and the two men began working on a problem soon after Heisenberg's return in the fall. It was during this time that Heisenberg met Albert Einstein for the first time. Einstein had been scheduled to talk on two previous occasions at Munich and Jena where Heisenberg and Pauli were both looking forward to meeting him. On both occasions, however, he had to decline because of anti-Semitism. Finally, in fall 1923 he came to Göttingen. Heisenberg talked to him for several minutes.

Heisenberg had received an invitation from Bohr to go to his Institute of Theoretical Physics in Copenhagen, and in March 1924 he went to Copenhagen for the first time. He was overwhelmed. "I was deeply depressed by the superiority of the young physicists from all over the world who surrounded Bohr," he later wrote. "Most of them could speak several foreign languages, while I could not express myself reasonably in even one . . . and they understood much more modern physics than I." Furthermore, many of them were excellent musicians and overshadowed Heisenberg's talents at the piano. This was dismaying to him since he knew he was an excellent pianist. Heisenberg was particularly envious of Bohr's assistant, Hendrik Kramers, who seemed to have an encyclopedic knowledge of physics.[10]

Heisenberg was mainly interested in talking to Bohr and soon became frustrated. Bohr was so busy for the first few days he had no time to talk. Finally, to Heisenberg's delight, Bohr suggested they go on a walking tour in northern Denmark, in a region called Zealand. Over the next few days, he and Bohr hiked through the countryside talking about physics and many other things. Bohr explained the history of the region to Heisenberg and asked him about the effect of the war on him. Much of the conversation, however, centered around the difficulties of modern atomic physics and how they could best be overcome.

Heisenberg was soon in awe of Bohr and his ideas. They were quite different from the views he had been used to in Göttingen and Munich. He came back to Göttingen with a fresh point of view. He also had an invitation from Bohr for a longer stay of several months. Later he referred to his talk with Bohr as "a gift from heaven."

Back in Göttingen, Heisenberg worked with Born on the helium atom. They now had a new approach and were sure it was going to succeed. But after several weeks Heisenberg realized it was hopeless. Both he and Born were becoming disenchanted with the current state of physics. "Basically we are convinced all present helium models are wrong, as is all of atomic physics," he wrote. Pauli also complained. Everyone was sure that something serious was needed, perhaps an entire revision of the basic principles.

On May 1, 1925, Heisenberg returned to Copenhagen for a longer stay. This time he decided he should learn some other languages. Everyone at the institute spoke several languages. Since he was in Denmark, it was important to learn Danish, but English was the major language used at the institute, so he felt he should also learn it. With the help of his landlady, he became conversant in both languages within a couple of months.[11]

During this stay he spent much more time with Bohr and learned a lot of new physics. More and more he began to appreciate Bohr's systematic, philosophical approach to physics. Mathematics was still important, but physics and understanding were emphasized. He also finally began to understand and see the crisis in physics much more clearly. While in Copenhagen, he began working on the problem of complex spectra with a new point of view in mind.

Heisenberg soon saw that dramatic changes would be required if he was to make progress. He finally decided to concentrate on the intensities of the spectral lines in hydrogen. They had never been predicted accurately. During the work he wrote to his father about his feeling of hopelessness for the future of physics. "Everyone here is doing something different, and no one anything worthwhile," he said in his letter. Everything seemed to be in such a muddle. Furthermore, Pauli had now proved his core model was incorrect. Then, to top it off, he found he couldn't work. He had an attack of hay fever that was so severe he could barely function. He had to get away from the pollen.

HELIGOLAND

In May 1925 he asked Born if he could have a couple of weeks off to recover and Born granted it. Off the coast of Germany was a tiny island known as Heligoland. It was rocky and barren—the perfect place to get rid of his hay fever. He took the train, then the ferry to the island.[12] Within a short time, his hay fever was gone and he was feeling much better. He could now concentrate on his problem. By now he had decided he could use the equations of the harmonic oscillator (actually, a slight modification of it) to approximate his problem.[13] Following Bohr's suggestion, he decided to forget about electron orbits around the atom. The only thing that was observable was the two energy states associated with the electron jump, or transition.

He began calculating transition rates between various states in the hydrogen atom. He was now dealing with arrays of numbers, so he set them up in a table. Strangely, though, he soon found that the variables in his calculations had an odd property. If, say, the position and momentum of a particle were multiplied in a certain order (i.e., $a \times b$), the product would not necessarily be the same if they were multiplied in the opposite order ($b \times a$). We refer to this as *commutativity*. He discovered that his variables did not commute. This bothered him, but he decided not to let it get in the way. As he continued calculating the transitions, he found that the numbers he was getting were agreeing exactly with known experimental values. He was immensely pleased.

It was very important, however, that energy was conserved. It was getting late as he made the calculations, but he was so excited he couldn't stop. He soon found that energy was conserved and the agreement with experimental values was excellent. When he finally lowered his pencil to the table, it was almost morning. At one end of the island was a large rock that he had wanted to climb. He went out and climbed it and watched the sun rise over the distant horizon.

AFTERWARD

He was excited about his results, but cautious. He stopped off at Hamburg on his way back to Göttingen to show them to Pauli. Pauli was always critical, and Heisenberg expected him to pick it apart and criticize it. To his surprise, however, Pauli was enthusiastic and encouraged him to continue. When he got back to Göttingen, he spent a couple of weeks writing up the results for publication. In the meantime he had received an invitation to speak at Cambridge University in England, and he decided to go.

He took his paper to Born to look over before he sent it in for publication, and then he headed for England. He was so exhausted by the effort he spent the first day in England sleeping. He slept so long several people thought he was sick.

THE NEW THEORY

Born looked over the paper Heisenberg had left with him. He was impressed with Heisenberg's ingenuity, but confused. The arrays of numbers that Heisenberg had used reminded him of something, but he couldn't remember what it was. Finally it came to him. As a student he had taken a class in advanced algebra, and in it he had been introduced to arrays of numbers called "matrices." They had the same properties as Heisenberg's arrays. In particular, they didn't necessarily commute.

Born was anxious to put the theory on a better footing, but he would need some assistance with the mathematics. He was describing his problem to a colleague on a train on the way to a scientific meeting when he was overheard by a mathematician from Göttingen, Pascual Jordan. Jordan was completely familiar with matrices and had worked with them extensively. He approached Born and offered to help. Over the next few weeks, the two men worked to put Heisenberg's ideas on a better footing, and by late summer they had succeeded. Heisenberg's paper was published in July, and Born and Jordan's paper followed in September.

MORE ELEGANCE

In England, Heisenberg lectured on several subjects, but he did not mention his breakthrough. In the audience was Paul Dirac. Dirac had started out as an electrical engineer, but after finding it difficult to find a job, he had switched to physics. He was working for Ralph Fowler, Ernest Rutherford's son-in-law.

Although Heisenberg never mentioned his breakthrough in his lectures, he did mention it to Fowler and told him that he was publishing a paper. Fowler asked for a reprint, and Heisenberg sent him one. Fowler read it and passed it on to Dirac. Dirac was not impressed, but as he studied it in more detail, he began to realize how important it was. He had a strong background in mathematics and soon saw that the theory could be reformulated. He made use of a classical theory by the Englishman William Hamilton and was able to put it into a much more elegant form. Heisenberg was very enthusiastic about Dirac's results. "I have read your extraordinarily beautiful paper on quantum mechanics with the greatest interest, and there can be no doubt that all your results are correct," he wrote to Dirac.[14]

Chapter Six

Schrödinger's Wave Equation

Heisenberg's new theory was a significant breakthrough, but not everyone was happy with his approach. Most physicists were unfamiliar with matrices. Furthermore, Heisenberg had taken the point of view that visualization was not important. Observations were the only things that were meaningful, and the only thing that could be observed was the transition between two states. The individual states, or orbits of the atom, took a backseat in Heisenberg's theory. Sommerfeld realized the importance of the new theory, but was less than enthusiastic about the approach Heisenberg had taken and the use of matrices. Einstein and Planck greeted the new theory with considerable skepticism. The view among many scientists was one of wait and see—perhaps a better formulation will come along, one that is more understandable and can be more easily visualized.

What most of them didn't expect, however, was that the new formulation would come so soon. Within a year of Heisenberg's discovery, what appeared to be an entirely different theory was formulated and published by Erwin Schrödinger of Zurich. It was based on differential equations, and therefore much more pleasing to most physicists. The differential equations of calculus were something they worked with all the time.

EARLY YEARS

Erwin Schrödinger was born on August 12, 1887, to Rudolf and Georgine Schrödinger of Vienna, Austria. Rudolf was a businessman who had taken over the family linoleum business. His heart, however, was never really in the business. Trained as a chemist, he had hoped to do chemical research, but had given it up. Throughout much of his life he regretted the decision. Schrödinger described his mother as "nice, with a cheerful character; she was poor of health and helpless towards life, but also unassuming."[1] She was, however, devoted to him. As the only child, Erwin was doted on by his mother and several aunts, along with several maids. He was surrounded by women throughout his youth, but his father still managed to have a strong influence on him.

Erwin Schrödinger

He was taught at home by tutors until he was eleven. His father believed this was better than sending him to elementary school and would give him a better chance of passing the entrance exams to the gymnasium. In 1898 he easily passed the entrance exams and was admitted to the gymnasium in the fall. Latin and Greek were emphasized at the gymnasium, but students also got a firm grounding in mathematics and science. From the beginning Schrödinger was an excellent student—always first in his class. Furthermore, he did it without studying hard. According to a fellow student, "[He] had a gift for understanding that allowed him without any homework to complete all of the material during the class hours and to apply it."[2]

Besides schoolwork he engaged in hiking, mountain climbing, and skiing. He also enjoyed the theater and liked poetry. Later in life he even wrote a book of poems. But strangely, he did not like music. This may have been inherited from his father, who also disliked music. This is strange in that almost all other well-known physicists of this time (for example, Einstein, Heisenberg, and Bohr) were music lovers. Schrödinger did, however, enjoy art.

When he graduated from the gymnasium, he looked forward to going to the University of Vienna. The physics department had a strong reputation. Many of the well-known early physicists were teachers there at one time. Ernst Mach, best known for his experiments on airflow and his discovery that sudden changes occur in the airflow at the speed of sound, had taught at the university for years. Ludwig Boltzmann, one of the most famous physicists in Europe, also taught there. He had done outstanding work on thermodynamics and had formulated the science of statistical mechanics (the branch of physics that used statistics and probability to deal with large numbers of similar particles). Schrödinger was looking forward to working under him. Unfortunately, in later life Boltzmann suffered from bouts of depression and just months before Schrödinger was to enter the university he committed suicide. Schrödinger was devastated when he heard the news. His only consolation was that he was able to study under Boltzmann's best student, Fritz Hasenöhrl.

Schrödinger entered the university in fall 1906. He had been a top student at the gymnasium and much was expected of him. He was already being thought of as a genius in mathematics and physics, and many of his fellow students were in awe of him. A friend, Hans Thirring, who entered the university a year later, recalled his first encounter with Schrödinger. Another student nudged him as Schrödinger entered the library. "It's the Schrödinger," the student said in a whisper.[3]

Schrödinger took theoretical physics from Fritz Hasenöhrl for eight semesters. The lectures included everything from mechanics to electrodynamics and thermodynamics. Hasenöhrl was young and enthusiastic, and he passed much of his enthusiasm on to his students. Schrödinger was particularly fond of him. "No other person had a stronger influence on me, except perhaps my father," he said. Indeed, Hasenöhrl was more than just a teacher. He frequently had his students over to his house, and he took them on hiking expeditions.

Schrödinger took experimental physics from the well-known experimentalist Franz Exner, who was also considered to be an excellent teacher. He particularly enjoyed his experience in Exner's lab. His mathematics professor, Wilhelm Wirtinger, was less inspiring; nevertheless, by the time Schrödinger graduated, he had an excellent grounding in mathematics. Indeed, his mathematical skills were exceeded by few in the field.

DOCTORATE

The first degree that Schrödinger received was a Ph.D. It is not considered to be equivalent to the present-day Ph.D., and was likely closer to an M.Sc. Besides the usual classes, he had to write a thesis. Despite his interest in theoretical physics, his thesis was on experimental physics. It was titled "On the Conductivity of Electricity On the Surface of Insulators in Moist Air." The work was performed under Professor Exner. What is surprising is that it contained literally no theory; in short, it was a straightforward set of electrical measurements. There were many interesting theoretical points related to the work that Schrödinger could have mentioned, but he didn't.[4]

Upon graduation he entered the military service on October 1, 1908. There were two options open to him. He could take the usual three years that were required or, as a university graduate, he could apply for officers' school. In this case only one year was required. He opted for the latter, and a year later had a commission as an artillery officer. He reentered civilian life on January 1, 1911, and was soon back at the University of Vienna on an assistantship.

Surprisingly, Schrödinger did not take his assistantship under Hasenöhrl. He took it in experimental physics under Exner. In later years he said that he never had any regrets. He felt that the experience he received in experimental physics was invaluable to him. He did conclude, however, that he was not cut out to be an experimentalist and should stick to theoretical physics. Despite this resolution, he continued to do experimental work off and on over most of his early career. The route to a professorship at this time was long and tedious. First came *habilitation*, which required several published papers, then you became a *privatdozent*, an unsalaried lecturer paid only through student fees. Even then the chances for a good position were not particularly good.

Schrödinger's first theoretical paper was presented to the Vienna Academy on June 20, 1912. It was titled "On the Kinetic Theory of Magnetism." It dealt with two forms of magnetism known as *diamagnetism* and *paramagnetism*. Other papers soon followed. His habilitation committee accepted his work in June 1913, with the comment, "It is the opinion of the committee that all the works of Schrödinger demonstrate a very well founded and broad scholarship and a significant, original talent." In October he took his oral exam, and on November 10 he completed his habilitation. He was now a privatdozent.

It was about this time that he almost quit physics. He became engaged to the daughter of some friends, but the young woman's mother was strongly against the marriage. There was no way he could get married on the meager wages of a privatdozent. Desperate, Schrödinger approached his father and told him he would like to quit the university and enter the family business so he could get married. His father said no, and urged him to stay at the university. He had regretted his decision to go into the family business all his life and was determined to make sure his son didn't make the same mistake. It is unlikely that the girl's mother would have allowed Schrödinger to marry her anyway. They broke up shortly after, and within a year he met the woman who would eventually become his wife.

In 1913 the Congress of Vienna was held at the University of Vienna. More than seven thousand people attended the meeting that included lectures on both scientific and medical topics. The featured speaker was Albert Einstein. His talk was titled "The Present Status of the Problem of Gravitation." Einstein had not yet developed his general theory of relativity, but his earlier contributions had made him one of the best-known theoretical physicists in the world. Schrödinger was impressed with Einstein's talk, and his interest in gravitation was aroused.

In March 1914 Schrödinger published his most important theoretical paper to date. It was published in the prestigious *Annalen der Physik* and had the title "On the Dynamics of Elastically Coupled Point Systems." It was based on the work of his idol, Ludwig Boltzmann.

WAR YEARS

Schrödinger was too busy with physics to take much interest in the politics of the day, but they were soon to effect him. With the assassination of Archduke Franz Fredinand, heir to the Austrian throne, most of Europe found itself embroiled in World War I. As a trained artillery officer, Schrödinger was called up immediately, so at the age of twenty-seven his physics career was on hold for several years. He was stationed at the Italian front. In many ways he was lucky in that, during the first part of the war, the Italian front was relatively quiet. He suffered, but more than anything, he suffered from boredom.

When the action became particularly intense, he and his fellow soldiers would head for a nearby grenade shelter. He wrote in his diary, "Losses [in the artillery] are like unlucky accidents. Normally nothing happens. Then all of a sudden a grenade finds a shelter or an observational station, and four or five people are torn to shreds. One should not have bad luck."[5]

Schrödinger fought at the battles of Isonzo where over 286,000 Italian soldiers and 140,000 Austrian soldiers were lost. At the end of the war, he was cited for outstanding service. Luckily, in 1917 before the war ended, he was shipped back to Vienna to teach meteorology to antiaircraft officers, so he was away from the fighting. During his spare time, he tried to do some research. As a theoretician all he really needed was a pencil and paper, and he did manage to get some work done. He published a paper on Brownian movement in July 1915 and another paper in July 1917 that dealt with the concussion and shock waves associated with large explosions.

During the latter part of the war, he was able to get a copy of Einstein's new theory of relativity. He immediately recognized its importance and became enthusiastic about gravitation, but it would be many years before he would try his hand at doing any research in the field.

Toward the end of the war, he became very despondent. He began to realize that his best years were being taken up by the war. In his diary he wrote, "I no longer ask when will the war be over? But will it [ever] be over?"[6]

POSTWAR LETDOWN

Finally, in 1918 he was released from service. He was thirty-one years old and starting to become pessimistic about physics. He applied for the chair of theoretical physics at the University of Czernowitz. By now he had decided to devote himself to philosophy. His beloved teacher Fritz Hasenöhrl had been killed early in the war, and out of respect for him he wanted to continue teaching theoretical physics, but Schrödinger was determined to do no research. He would devote his time to philosophy. As it turned out, nothing came of the job and he had to stick with physics.

The years immediately after the war were hard ones for Schrödinger. His father's business had failed and the family was broke. Then on Christmas Eve 1919 his father died. He was sixty-two. Schrödinger was devastated; he had always been close to his father. A little later his grandfather and his mother died. To make things worse, Austria had been blockaded after the war, there were extreme shortages, and inflation was rampant. The only bright spot was Anny Bertel. She had visited him during the war, and they were now talking of marriage. Although he was uncertain he could support a wife, he and Anny were married on March 24, 1920.

Soon Schrödinger got a position at the University of Jena where he worked with Willy Wien. He was becoming interested in atomic theory and gave his inaugural lecture on the topic. To his surprise, he had been there only a few weeks when he was promoted to associate professor. But inflation was still out of control, and his salary was far from adequate for a decent living. So when he was offered an associate professorship at Stuttgart in October 1920 with a considerably higher wage, he took it.

While at Stuttgart, Schrödinger received a copy of Arnold Sommerfeld's book *Atomic Structure and Spectral Lines*. It was a complete compendium of Bohr's theory along with Sommerfeld's contributions to the theory, and it soon became the most important book in the area. Schrödinger studied it in detail and was so pleased with it he wrote to Sommerfeld: "You are really an unselfish teacher to all of us, using your strong teaching abilities to enable as many as possible to elaborate further along the path you have prepared."[7]

Schrödinger's stay in Stuttgart was short. A position at the University of Breslau became available, and Schrödinger applied for it. He had second thoughts about moving to Breslau because it was close to the Polish border, but they offered him a full professorship and he accepted it.

Conditions in both Austria and Germany were appalling. Food and fuel were in short supply, and the currency was worthless. Even though Schrödinger was now a full professor, his living conditions were still far from adequate. Before he came to Breslau, he had been asked if he would accept a professorship at Zurich, Switzerland, if it were offered to him. Conditions in Switzerland were much better than those in Germany, so Schrödinger said that he would. There was, however, considerable uncertainty about the position. Officials at Zurich had written Sommerfeld asking for a list of potential candidates for the position. Sommerfeld had given them a list of six people. Schrödinger was one of them, but he was not first on the list. About Schrödinger he wrote, "A first-rate head, very sound and critical. He is a full professor at Breslau and thus certainly not available to you [as an associate professor]."[8]

When the vote was taken at Zurich, Schrödinger was not the first choice. But their first choice turned the offer down, and one of the other men on the list accepted another position. The offer was finally made to Schrödinger, and he accepted.

ZURICH

The Schrödingers arrived in Zurich in October 1921, and Erwin was exhausted. He had barely started lecturing when he had a serious case of bronchitis, and it appeared that he might have tuberculosis. The doctor ordered him to take a long rest at a high altitude. Arosa, an alpine resort near a ski area, was selected, and for the next nine months he rested. He did, however, manage to write two short papers on specific heat. Finally, at the beginning of the winter semester of 1922–23, he was back at work with a relatively heavy teaching schedule. He soon became best friends with the brilliant mathematical physicist Herman Weyl, who had just written *Space-Time-Matter*.[9]

Conditions were much better in Switzerland than they had been in Germany. "If I imagined having to go back to Germany, I'd be horrified," he wrote to a friend. He was happier than he had been in some time, but he was beginning to worry. He was now thirty-seven years old and had worked hard in so many different fields of physics, but had never done anything of real significance. He was all too familiar with the old saying, "Once you're over thirty, it's too late." Heisenberg, Bohr, Einstein, and others had done their most creative and brilliant work before they were thirty. Actually, Einstein did some of his best work when he was over thirty, but he appeared to be an exception. Schrödinger knew that time was running out. He was a respected physicist, but not considered to be very creative. He had been invited to the prestigious Solvay Congress in Brussels along with the best physicists of Europe, but he was not asked to present a paper and was merely an observer.

As the months passed, things seemed to deteriorate. He was now fighting with his wife, Anny, and there was talk of a divorce. He had wanted children and she couldn't have any. He got an offer from Innesbruck, Austria, and was tempted to accept it. He had always loved Austria, but he knew that conditions there were still poor, and he turned the offer down.

THE ETH–UNIVERSITY OF ZURICH COLLOQUIUM

In Zurich there were two universities: the University of Zurich and ETH (Zurich Polytechnic). Einstein had earlier been associated with both of them. He had graduated from the ETH and had got his doctorate from the University of Zurich. Later he had taught at both institutions. Neither institution had a large physics faculty, so they joined together for their colloquium. This colloquium was led by Peter Debye of ETH, and it met every two weeks. Schrödinger, of course, participated fully in it.

In fall 1925 Schrödinger was lecturing on molecular statistics and had begun doing some research in the area when he came across a paper by Einstein in which Einstein referred to de Broglie's thesis. Schrödinger wrote to Einstein asking about the thesis. In a second letter to Einstein a few weeks later, he wrote, "A few days ago I read with the greatest interest the ingenious thesis of de Broglie, which I finally got a hold of; with it section 8 of your second paper becomes clear to me for the first time." Soon afterward Schrödinger published a paper related to Einstein's work.[10]

Meanwhile, by late November there was considerable excitement about de Broglie's work and some talk of it at the colloquium. At the end of one of the colloquia, Debye said to Schrödinger, "Schrödinger, you're not doing anything important now, why don't you give us a talk on de Broglie's work." Schrödinger accepted the invitation happily since he had already begun studying the theory.[11] In the next colloquium, he gave a clear account of de Broglie's work, describing his standing waves and how they accounted for Bohr's orbits. When he finished, Debye stood up and said, "Schrödinger, you are talking foolishly. You are talking of waves, but you have no wave equation."[12]

Felix Bloch, who was at the meeting, later said, "The remark sounded quite trivial and did not seem to make a great impression, but Schrödinger evidently thought a bit more about the idea." Schrödinger did, indeed, think about the idea, and within a short time he began looking for a wave equation. De Broglie had presented his theory in a relativistic form, so Schrödinger naturally looked for a relativistic form of the wave equation, and within a short time he had one. When he compared its predictions against experimental results, however, he was disappointed. It gave incorrect results.

Schrödinger's relativistic wave equation was actually correct. Why it appeared to give incorrect results was that it automatically incorporated the spin of the electron. In other words, it gave the results for a spinning electron, and spin had not been discovered at that time. Schrödinger put the problem aside. The semester was nearing an end, and he would have some time during the Christmas holidays to look at it again.

THE BREAKTHROUGH

As Christmas break approached, Schrödinger began to think about the nonrelativistic case. If he neglected relativity, he would also get a wave equation, and it would be valid in all cases except extremely high speeds. His life was in considerable turmoil at this time. He had been fighting continuously with Anny for weeks, and he knew he had to get away if he was to get any work done. He decided to go to Arosa, where he had recovered from his tuberculosis. He wrote an old girlfriend to meet him there in mid-December. Little did he know that the next two weeks would be the most important of his life and would create one of the most important upheavals in physics. It has been said that if Schrödinger had died at thirty-eight, just before his stay at Arosa, he would have been little more than a footnote in the history of physics.

According to Herman Weyl, "Schrödinger did his great work during a late erotic outburst in his life." Indeed, despite considerable searching, we still don't know the identity of the woman with whom he spent the two weeks. According to Walter Moore, "During this time he began a twelve month period of sustained creative activity that is without parallel in the history of physics." He took a break on December 27 and wrote to Willy Wien at Munich: "At the moment I am struggling with a new atomic theory. If only I knew more mathematics! I am very optimistic about this theory and expect that if I can only . . . solve it, it will be very beautiful."[13]

Schrödinger soon had an equation that he applied to the hydrogen atom. Surprisingly, he had still not solved it when he arrived back in Zurich on January 9, so he went to his friend Herman Weyl to ask about the solution. Weyl gave him several hints, and he was soon able to solve it. Back at the university, one of the deans asked him if he had enjoyed the skiing over his vacation. He said that he hadn't done any skiing—he had been distracted with some calculations.

Soon after the beginning of the semester, Schrödinger gave his second colloquium on de Broglie's work. He began his lecture by saying, "Professor Debye suggested that one should have a wave equation. Well, I have found one." It was soon evident that he had made a major discovery.

FIRST PAPER ON WAVE MECHANICS

Schrödinger's first paper on *wave mechanics* was received by *Annalen der Physik* on January 27, 1926. It was titled "Quantization as an Eigenvalue Problem." Schrödinger had obtained his wave equation by substituting de Broglie's condition for the wavelength of a particle into the well-known classical wave equation. Worried that this might appear too easy, and unconvincing, he gave a derivation of it in his first paper. Since there is no such thing as a logical derivation of a quantum equation from a classical one, his derivation was more of a justification than an actual derivation.

The solution for the case of the hydrogen atom gave the same results as Bohr's theory, but Schrödinger now had a powerful weapon: an equation that could be applied to any problem in atomic physics. One of the difficulties of the new theory, however, was what is called the wave function ψ. The wave equation is, in fact, an equation for ψ (pronounced as psi), but what did ψ represent? Schrödinger was uncertain, and for some time this was a problem.

THE SECOND AND THIRD PAPERS

Schrödinger didn't rest after publishing his first paper. The equation he had derived could be applied to many problems, and over the next several months he worked diligently to solve many of them. His second paper was received by *Annalen der Physik* on February 23, 1926. It included a new derivation of the wave equation and an application of the equation to the harmonic oscillator and the diatomic molecule (a molecule with two atoms). One of the most important results of the paper was that he showed that the wave function ψ depended on space coordinates. Schrödinger mentioned in the paper that he was aware of Heisenberg's new theory, but he had no idea how it related to his.

In the third paper, which was received by *Annalen der Physik* on May 10, 1926, Schrödinger dealt with the Stark effect, in other words, the splitting of spectral lines when the atom is in an electric field. It was an extensive paper—fifty-three pages long. In it he used perturbation theory for the first time. This is a technique where a series of approximations (first order, second order, and so on) are made to the exact solution. It is a powerful tool in all complicated problems.

By now the reactions to his first papers were beginning to pour in. Planck wrote that he had read the first paper "like an eager child having the solution to a riddle that had plagued him for a long time." After reading the second paper, he wrote, "You can imagine with what interest and enthusiasm I plunged into the study of his epoch making work." Planck brought the papers to the attention of Einstein, and within a short time Einstein wrote to Schrödinger, "The idea of your work springs from true genius."[14] Schrödinger was particularly touched by this comment. Paul Ehrenfest of Leiden wrote, "I am fascinated by the . . . theory."

THE LAST PAPER

The last of his six papers was received by *Annalen der Physik* on June 23. In his previous papers he had dealt only with stationary systems, in other words, systems that did not change in time. In this paper he dealt with systems that do change in time. The most important of these problems are those relating to the absorption and emission of radiation by atoms, and the scattering of radiation by atoms. This paper showed Schrödinger's tremendous insight into the problems and his true genius. It was, indeed, an epoch-making paper.

There was so much demand for Schrödinger's papers he brought them all together in a book in November 1926. For many years it was the "bible" for people interested in learning the new theory. In the foreword he wrote, "A young friend of mine recently said to [me]: 'Hey, you never even thought when you began that so much sensible stuff would come of it.' " This was no doubt quite true.

Looking back over Schrödinger's career, it's amazing how many things could have changed his fate. Wave mechanics would likely have been discovered in time, but it wouldn't have been Schrödinger that discovered it if any of several things had happened. Just after he graduated, he wanted to abandon his career and join the family business so he could get married. But his father would have none of it. Later he decided to abandon physics and go into philosophy. This was going to occur after he got the job at

Czernowitz. And even just before he made the discovery, he was seriously considering leaving Zurich and taking a job at Innesbruck, but he finally decided against it.

Of course none of these happened and Schrödinger did, indeed, discover the equation that now bears his name and the branch of physics now known as *wave mechanics*. And as Sommerfeld said a few years later, "It was the most astonishing among all of the astonishing discoveries of the 20th century."

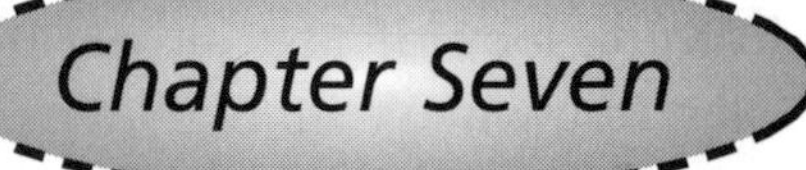

What Did It All Mean?

Scientists now had two theories of the atom. They were radically different in nature, but they solved the same problems. When applied to such things as the harmonic oscillator, the hydrogen atom, and the Stark effect, they gave exactly the same answer. What was so strange was that not only were the two theories based on entirely different concepts but their philosophical implications were also different. Schrödinger's theory was visual in that it allowed you to picture what was going on. His "standing electron waves" were representations, albeit somewhat fuzzy, of Bohr's orbits, but his theory was free of the ad hoc postulates of Bohr's theory, and it was not limited in the way Bohr's was.

Also, to the delight of many, Schrödinger's theory eliminated the need for "jumps" between orbits. Transitions, of course, occurred, but they were from one standing wave to another. They were like the transitions of a standing wave on a stretched string. If you attached one end of the string and moved the other end appropriately, you could set up a standing wave that consisted of one loop. If you moved it slightly faster, you could produce a standing wave with two loops. The change from one state to the other was continuous, and Schrödinger's theory was therefore thought of as a continuous theory. Its most pleasing feature, as far as most scientists were concerned, was that it was based on differential equations rather than matrices, and differential equations were an everyday tool of all physicists. Matrices, on the other hand, were foreign to most physicists. Another feature that was pleasing to many was that there was a direct link to classical theory. Many of the concepts in Schrödinger's theory were taken from classical theory, including the wave equation itself.

Heisenberg's theory, in contrast, was based on discontinuities, and quantum jumps were an intricate part of the theory. It was a mathematical scheme that did not allow visualization and had no real connection with classical mechanics.

REACTION TO SCHRÖDINGER'S THEORY

The response to Schrödinger's theory was immediate, and it came from many directions and places. Most of it was positive, and it is easy to see why. Problems that had resisted solution for years were now suddenly solvable. Few criticized the mathematics, but there was criticism of the implications of the theory. Nevertheless, it was greeted with tremendous enthusiasm. Hartmut Kallman of the University of Berlin gives us some idea of the enthusiasm. He reported, "The colloquium, which was usually attended by 20, was attended by 200 when the new theories were discussed." He went on to say that the halls were crammed with so many people that many of them could not get into the lecture hall.[1]

The correspondence of the day also shows the enthusiasm. Wolfgang Pauli wrote to Pascual Jordan, "I feel this paper is to be counted among the most important recent publications. Please read it carefully and with devotion."

Arnold Sommerfeld wrote to Pauli, "A manuscript by

Schrödinger for *Allanen* has arrived. Schrödinger seems to obtain exactly the same results as Heisenberg and you, but in a quite different, totally crazy way . . . by boundary value problems. Surely something reasonable and fundamental will soon emerge from all that."[2]

Neils Bohr wrote to Heisenberg, "[Pauli] is fascinated by the first published paper of Schrödinger."

The theory ignited a firestorm in Germany and many of the neighboring countries, but strangely there was little reaction from France, where in a sense it all started with Louie de Broglie. De Broglie did not take advantage of the theory and added little to it, perhaps because he was so particle oriented. To him, particles had associated waves, but they were not made up of waves. There was also little response in England initially, except for Paul Dirac's contribution. The response from America was positive, and there was considerable interest at most major universities, but few contributions to the theory.

Once the initial flurry had subsided, the question remained: Why were there two theories for the same phenomena? Was it possible they were connected?

HEISENBERG AND SCHRÖDINGER

Heisenberg was not pleased with Schrödinger's theory when he first heard of it. He wrote to Pauli, "The more I ponder the physical parts of Schrödinger's theory, the more horrible I find it." His major criticism, however, was directed at the implication of the theory, not the mathematics.[3]

The animosity was not in one direction. Schrödinger stated several times that he had little use for Heisenberg's nonvisual theory. He hoped that once his new theory became established "matrices would go away." In a letter he wrote, "I was discouraged, if not repelled, by what appeared to me as a very difficult nettle of transendental algebra, defying any visualization."

On the other hand, according to Heisenberg, "[Schrödinger's theory] does not lead to a consistent wave theory in the de Broglie sense." And in a letter he wrote to Pauli, "[Schrödinger's theory] does not fit the facts. . . . It does not solve problems that require discontinuity."

Pauli mentioned this fact to Schrödinger, emphasizing that a discontinuous element somehow had to be introduced into the theory if it was to give a true understanding of quantum phenomena. Schrödinger felt that this would come as the theory was better understood.

As Heisenberg learned more about the theory, he continued his bombardment. In a letter to Jordan, he wrote, "I am rock-solid convinced of the inconsistencies of the physical interpretation of the quantum mechanics presented by Schrödinger." And in another letter, "[Schrödinger's] purely continuous description of the experiments in the sense of classical theory simply contradicts accepted results."[4]

With such a strong reaction, it might seem that the animosity between the two men may have gotten out of hand, but apparently it didn't. It was not personal. In one of his letters, Heisenberg wrote, "As nice as Schrödinger is personally, . . ."

Despite Heisenberg's severe initial criticism, most physicists, including Einstein and Planck, preferred Schrödinger's approach. But many, including Schrödinger, soon became convinced there had to be a connection between the two theories. How could they give *exactly* the same results and not be related?

THE CONNECTION

Soon after he published his first paper, Schrödinger began to wonder about a possible relationship between his and Heisenberg's theories. And he wasn't the only one. Sommerfeld, in Munich, was also convinced that the two theories had to be related. He relayed his concern to Wien, who wrote Schrödinger about the matter. Soon after he received the letter, Schrödinger sat down and carefully compared the two theories, but he quickly became discouraged. In a letter to Wien, he said, "In relation to

Heisenberg's [theory], I am convinced along with Sommerfeld that an intimate relation exists. It must, however, lie rather deeply. I have given up looking any further myself."[5]

Within a short time, however, Schrödinger must have taken another hard look, for two weeks later Wien received a letter from him that began, "The hope I expressed in my letter of February 22 has been fulfilled earlier than I thought. The relationship to Heisenberg's theory has now been completely clarified."[6]

Schrödinger published his result in the March 18, 1926, *Allanen Der Physik*. What Schrödinger did was show that Heisenberg's matrices could be constructed from the solutions of his differential equations. These solutions are referred to as *eigenfunctions*, so what Schrödinger did, in essence, was show that a series of eigenfunctions gave the components of Heisenberg's matrices.[7] He also showed that the relationship was reciprocal; in other words, Heisenberg's matrices gave his eigenfunctions.

Interestingly, Schrödinger barely beat several others to the solution of this problem. Pauli was already working on it, and shortly after Schrödinger obtained his result, Pauli arrived at the same result. Furthermore, Carl Eckhart in the United States also showed the equivalence of the two methods. His result appeared in *Physical Review*, but it did not appear until several months after Schrödinger's paper. Finally, it soon became evident from Dirac's transformation theory that the two approaches had to be equivalent.

PHYSICAL INTERPRETATION

With the problem of the relationship of the two theories out of the way, most people turned to another problem: the meaning of various elements within Schrödinger's theory. In particular, what was the significance of his wave function? The wave equation gave both eigenfunctions and eigenvalues. There was no problem with the eigenvalues; they were observables such as energy and momentum. But what did the eigenfunctions, or wave functions, represent?

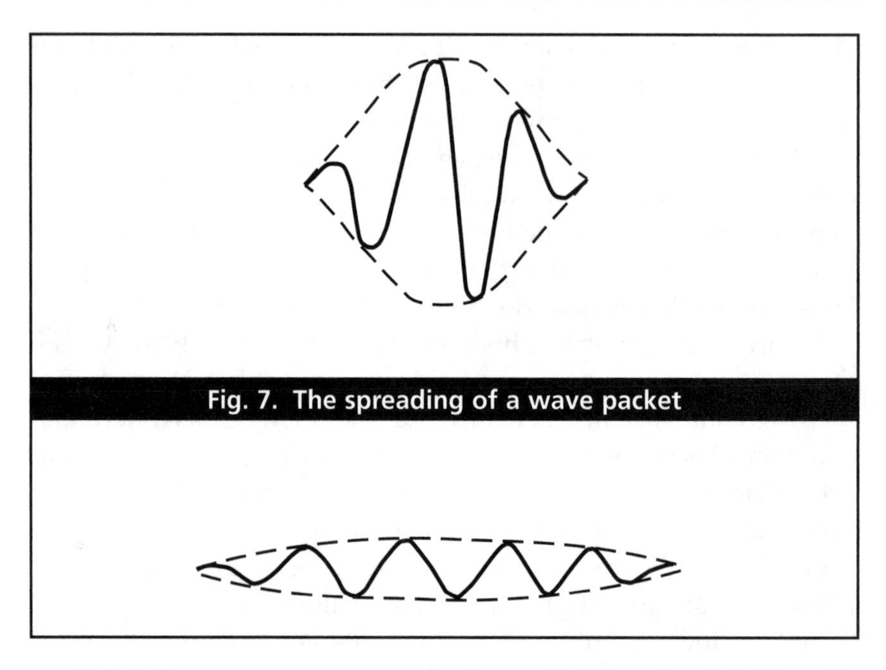

Fig. 7. The spreading of a wave packet

Schrödinger came to a conclusion early. Shortly after he published his first paper, he became convinced that the wave functions represented the matter of the particle. In other words, the particle was made up of waves, and these waves gave a measure of the matter, or more exactly, the distribution of the matter of the particle. Furthermore, he showed that many waves could be superimposed, just as sound waves of music can be superimposed, and from this superposition he introduced the idea of a *wave packet*. According to him, this wave packet represented the particle. This appeared to make sense. A moving particle would then be a *pulse*, like the pulse that moves down a tightened string. But there was a serious difficulty. A wave packet was composed of many waves of different frequencies and wavelengths, and the individual waves traveled at different velocities. This meant that when the packet moved it would spread apart (fig. 7). In fact, the wave packet for the free electron would spread out to the size of a house in a millionth of a second. Shazam! and it would be gone.[8]

But electrons in nature did not do this. It was easy to follow the track of an electron in a Wilson cloud chamber, and it did not spread out as it moved. Physically, therefore, the electron couldn't be a wave, or a wave packet. But if it wasn't a wave packet, what was it? and what was the significance of the wave function?

Schrödinger and others worried about this problem, but Schrödinger stuck to his guns. He was convinced that particles were waves or made up of waves. It seemed to be the only reasonable interpretation.

BORN'S INTERPRETATION

Max Born was not satisfied with Schrödinger's interpretation. He was closer to experiments involving electrons than some of the other theoreticians and found the concept hard to accept. His colleague James Frank and others at Göttingen had been bombarding various gases with beams of electrons, and there was no evidence in their experiments that the electrons were expanding wave packets. After considerable thought he finally arrived at an interpretation: the wave functions gave a measure of the *probability* of the position of the particle. More exactly, the square of the wave function gave the probability that the particle was at a certain point.

Every wave had an amplitude, which was the height of the wave. If the amplitude of the wave was high, the probability that the electron was at this point was also high. If the amplitude was low, the probability that the particle was at this point was low. Born published his paper in spring 1926. He was pleased with the new interpretation, and was convinced it was correct, but his paper got a mixed reception.

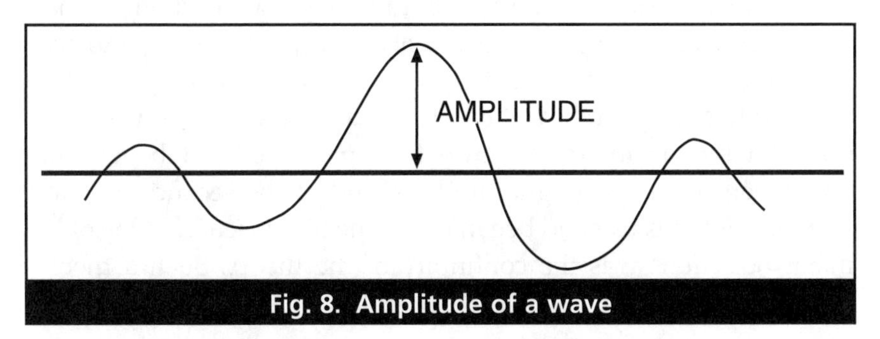

Fig. 8. Amplitude of a wave

According to Born, Schrödinger's cloudlike patterns for the electrons in an atom were not matter clouds. They gave the probability distribution for the position of the electron and could not tell you exactly where the electron was at a given time. Bohr and Heisenberg eventually accepted this interpretation, but Schrödinger did not. He continued to stick with his own interpretation.[9]

THE AFTERMATH

Schrödinger's life changed dramatically after he published his theory. Everyone was after him to talk about his new theory, and his life became a whirlwind of activity. Max Planck at the University of Berlin repeatedly invited him to talk in Berlin, and Sommerfeld and Wien were after him to come to the University of Munich. In midsummer he decided to visit both universities. He arrived in Berlin on July 16 and was cordially met by Planck and others. His talk, as usual, was clear and elegant. Einstein was in the audience and was impressed with the new theory. Later, however, he would have reservations. Not only did Schrödinger discuss the successes of the theory, but he also talked about its shortcomings. He mentioned that a relativistic version of the theory had not yet been discovered and there were problems with applying the theory to certain magnetic phenomena. The concept of *spin* had just been discovered by George Uhlenbeck and Samuel Goudsmit, and Schrödinger was sure that when it was incorporated in the theory these problems would be overcome.

On July 23 Schrödinger traveled to Munich. He was reunited with Wilhelm Wien and Arnold Sommerfeld, with whom he had carried on an extensive correspondence. Everyone was eager to hear about the new theory. Heisenberg was now at Copenhagen, but he traveled to Munich to hear the lecture. Later he wrote, "I was really horrified by the theory, because I simply could not believe it." Heisenberg held his tongue until the end of the second talk, and then he got to his feet and began to outline his objections. One of his major objections was the continuity of the theory. Such a theory, according to him, would never be able to explain discontinuous phenomena such as the photoelectric effect, the Frank-Hertz experiment, or the Stern-Gerlach experiment on spin. As Heisenberg talked it soon became evident that few in the audience were agreeing with him. Several people objected to his arguments, especially Wien, who got to his feet and told him to sit down and be quiet.

Heisenberg left the meeting depressed and distraught. He hadn't been allowed to finish his arguments, and sentiment was obviously strongly against him. He went hiking with some of his friends for several days in the surrounding mountains as he thought over his objections. At the time he was working on the helium problem using his own matrix theory, and he soon realized that some of the calculations would be easier using Schrödinger's theory. He published his results shortly thereafter.

VISIT TO COPENHAGEN

Bohr had also been after Schrödinger to talk in Copenhagen. Schrödinger accepted the invitation and arrived in Copenhagen in early October 1926.[10] Bohr was looking forward to the visit; he had studied Schrödinger's theory thoroughly, and he was familiar with Born's interpretation of the waves as probability waves, but his main interest was the philosophical implications of the theory.

Heisenberg was also now back in Copenhagen, and, as he related later, the arguments started almost from the moment Schrödinger arrived. He had barely got off the train when Bohr began bombarding him with questions. The discussions and arguments continued from early morning until nightfall for every one of the days Schrödinger was in Copenhagen. As Heisenberg said, Bohr, who was usually kind and considerate of people, was now like a "relentless fanatic who was not prepared to concede a single point."

The arguments centered mainly on two main topics: electron jumps and Born's probability interpretation. One of the things Schrödinger had been particularly pleased with when he discovered his theory was that electron jumps from one orbit to another were not needed in his theory. The state of a particle was represented by a wave cloud, and when a transition occurred the cloud would merely "fade away" and be replaced by another cloud. It was a smooth, continuous transition, and no electron jumps were required. Bohr disagreed. Furthermore, he objected strongly to Schrödinger's interpretation of the wave cloud as a matter cloud. By now he was convinced that Born's probability interpretation was the correct one.

The arguments continued with neither man giving an inch. Bohr had been sure he could convince Schrödinger, but he made little progress. Heisenberg, as Bohr's assistant, no doubt also joined in on some of the discussions, but it was primarily Bohr and Schrödinger going head-to-head. At one point Schrödinger said, "If we are still going to have to put up with those damn quantum jumps, I am sorry that I ever had anything to do with quantum theory."[11]

Bohr replied, "But the rest of us are very thankful for it—that you have—and your wave mechanics with its mathematical clarity and simplicity is a gigantic progress over the previous form of quantum mechanics."

Schrödinger finally fell sick and had to take to bed, no doubt as a result of the stress. He was looked after by Mrs. Bohr, but this didn't stop Bohr. He sat on the edge of the bed and continued the arguments.

In the end there was no agreement. In later years Schrödinger said that Bohr almost convinced him, but there was no indication of it at the time. The main reason, no doubt, was that Schrödinger felt strongly that visualization was important, and Bohr did not.

After his final talk to the Danish Physical Society, Schrödinger left. He was impressed with Bohr and glad he had met him. (This was, in fact, the first time the two men had met.) But he was relieved to leave Copenhagen since it had been very stressful.

TRIP TO AMERICA

Schrödinger had barely got home from his travels throughout Europe when he received an invitation from the United States. His theory had attracted considerable attention there, and many were anxious to hear him talk. The invitation came from the University of Wisconsin at Madison, but he would be talking at many of the major universities. On December 18, 1926, he left for the United States on a French liner.[12]

The voyage was not a pleasant one for him. Anny was seasick most of the way over, and his first impression of the United States was less than favorable. Of the eastern cities, only Madison appealed to him. He was offered positions at the University of Wisconsin and Johns Hopkins University, but he turned down both mainly because he had heard earlier that a position at the University of Berlin would soon be available and he would be a leading candidate.

From Madison he traveled to California where he lectured at the California Institute of Technology and several other universities. He was taken up the long, twisting road to the top of Mount Wilson to the then largest telescope in the world, the 100-inch Hooker reflector. In all he gave fifty-seven lectures in the United States and was exhausted by the time the trip ended. He arrived back in Europe on April 10.

PLANCK'S CHAIR AT BERLIN

Max Planck was nearing retirement age, and this meant that his chair of theoretical physics at the University of Berlin would soon be vacant. It was one of the most prestigious physics positions in Europe. Schrödinger was being seriously considered for it, but he was not the only candidate. The committee's first choice was Arnold Sommerfeld of the University of Munich. Others on the list were Max Born of Göttingen, Peiter Debye of ETH in Switzerland, and Werner Heisenberg. Although Heisenberg had made an important breakthrough in atomic physics, he was only twenty-four and generally considered to be too young for the position.

After considerable debate Debye was eliminated because he was primarily an experimentalist, and it didn't seem appropriate to offer him a chair in theoretical physics. Sommerfeld turned down the offer, so only Born and Schrödinger were left on the list. The offer was finally made to Schrödinger in summer 1927. Despite the prestige of the chair, Schrödinger was reluctant to take it at first. He hated to leave Switzerland, and he dreaded the politics of Prussian society. Furthermore, Zurich did not want to lose him. They could not match the wages Berlin was offering, so they got together with officials from ETH and made him a joint offer that was close to that of Berlin. The problem with their offer was that he would now have to lecture at both institutions, and his teaching load would increase considerably. His teaching load at the University of Berlin would be minimal.

Schrödinger finally accepted the offer from Berlin, and he and Anny moved in August 1927.

THE FIFTH SOLVAY CONFERENCE

One of Schrödinger's first functions after accepting the position at the University of Berlin was attending the fifth Solvay Conference which took place in Brussels from October 24 to 29.[13] Schrödinger had attended the previous conference, but only as an observer. The situation was quite different this time; he was now recognized as one of the leaders in theoretical physics, and everyone was eager to hear his talk.

The talks were held in one of the leading hotels in Brussels. Schrödinger, who never cared for formal wear, arrived at the hotel in one of his Tyrolean costumes with a knapsack on his back. Dirac, who met him in the lobby, took one look at him and said, "Schrödinger, you look like a bum from the streets." And indeed, the photograph of the participants shows Schrödinger in a light checkered sports jacket and bow tie, with everyone else in dark suits and high collars. Even Einstein, who generally detested formal wear, wore a dark suit and a tie.

Schrödinger's talk was, of course, one of the highlights of the conference. The interpretation of his wave function was still controversial, and although he briefly mentioned his version, he said he would leave any discussion of the probability interpretation to Born and Heisenberg, who were in the audience. They also gave talks at the conference.

One of the highlights of the conference was the heated discussions between Einstein and Bohr about the philosophical implication and interpretation of quantum physics. This went on for several days, and it brought several important points of interpretation into the open that will be discussed in detail later.

LATER YEARS

Schrödinger was skeptical when he got to Berlin, but to his surprise both he and Anny enjoyed the city. Schrödinger loved the theater, and he now had his choice of theaters. In July 1929 he was inducted into the prestigious Prussian Academy of Science. During his stay in Berlin, he became good friends with Einstein; both men looked forward to the weekly colloquium where new discoveries and theories were discussed at length. Many interesting discussions occurred at these meetings.

There was, however, considerable political turmoil in the city. The Nazi Party was increasing in strength, and in January 1933 Adolf Hitler was appointed chancellor of Germany. Schrödinger disliked the Nazis, and when they began persecuting the Jews, he decided to leave Berlin. Without formally resigning his position, he left and went to England where for several years he taught at Oxford University. He had barely arrived at Oxford when he was informed that he was to be awarded the Nobel Prize. He had been nominated by several people in 1929 and 1930 but had not won. Strangely, no prize was awarded in the years 1931 and 1932. The committee finally decided in 1933 to award the 1932 prize to

Heisenberg and the 1933 prize to Schrödinger and Dirac jointly. All three prizes were actually awarded at the same time.

Schrödinger returned to Graz, Austria, in 1936, but the Nazis soon annexed Austria and he was under Hitler's regime again. Three years later he fled to Dublin, Ireland, where he spent the next seventeen years.

Although Schrödinger probably didn't realize it at the time, the birth of his theory would do much more than introduce a new technique for dealing with atomic problems. With Born's interpretation of probability waves, the determinism of the old physics was gone—replaced by indeterminism, and this, in turn, would lead to many strange implications.

Chapter Eight

Uncertainty

As Bohr's assistant, Heisenberg was involved in the discussions and arguments between Schrödinger and Bohr at Copenhagen, and they left a strong impression on him. Despite Schrödinger's stubbornness, both men felt that his point of view had been adequately refuted. Heisenberg had been committed to the particle aspect of quantum theory with its discontinuities and quantum jumps, and he was even more committed to it after Schrödinger's visit. Schrödinger's theory, with its waves, was a continuous theory, and the way things stood it could not describe discontinuities. Heisenberg was also beginning to be bothered by the tremendous interest in Schrödinger's theory and the sudden lack of interest in his own theory. Many papers were now appearing on wave mechanics, and the number employing matrix mechanics was dwindling. It particularly bothered Heisenberg that some of the papers on wave mechanics were nothing more than the reworking of problems that had already been solved using matrix mechanics.

After Schrödinger left

Copenhagen the discussions and arguments did not subside. As far as Heisenberg and Bohr were concerned, many serious problems remained, and most of them were related to the philosophical implications of the theories. Although the two theories had been shown to be equivalent, both men hoped that a better unification would be found.

Heisenberg was living in a room in the attic of Bohr's institute, and Bohr would frequently come up late in the evening to discuss the problems. Occasionally these discussions would last well into the night. On one particular occasion, Heisenberg was so wound up he knew he would never sleep; he had to relax and clear his mind, so he went for a walk.[1] It was clear and the stars shone brightly. As he looked up at them, he thought about what he and Bohr had discussed. Suddenly he realized something important was being left out. There had been little discussion of the measurement process and how nature should be observed, yet it was important, and it had a strong bearing on the theory.

THE UNCERTAINTY PRINCIPLE

Heisenberg began to think seriously about the measurement process. There were four main variables that were measured: position, momentum, energy, and time. Position and momentum were intimately related, as were time and energy. We refer to these pairs as being *conjugate*. Heisenberg knew that if quantum theory was to make sense, you had to be able to relate it to observation and the measurement process. He thought about the process in relation to the electron. Was it possible to measure the position and momentum (speed) of an electron accurately at the same time? After thinking about it for a while, he realized it wasn't. To measure the position of the electron accurately, you would need a microscope. Furthermore, because the electron was so small, this could not be an ordinary microscope. For high accuracy you would need very short wavelength light. In fact, ordinary light would have wavelengths that were too long. A gamma-ray microscope would be needed.

It is perhaps ironic that he was thinking about a microscope and how it could be used to observe particles. He had almost flunked his doctorate oral because he didn't know anything about microscopes. Wien had asked him to derive an expression for the resolving power of a microscope, and he didn't know how to do it. He was so embarrassed, however, that after the exam he sat down and learned everything there was to know about microscopes, not realizing that one day it would be invaluable to him.

As he thought about the experiment with the microscope now, he realized the knowledge he had acquired would be helpful. It was obvious to him that if you decreased the wavelength so that you could see the electron better, you were, at the same time, increasing its energy. The energy of radiation depends on its wavelength, with the shortest wavelengths having the highest energy, and gamma rays, which would be needed to see the electron clearly, had tremendous energy. This meant that when a gamma-ray photon struck the electron it would impart considerable energy to it, causing the electron to recoil vigorously. Because of this, it would be difficult to measure the momentum (or speed) of the electron accurately. The only way you could get an accurate measurement was to increase the wavelength of the light, so it wouldn't be so energetic. And, of course, it's easy to see there's a catch-22 here: when we increase the wavelength so we can measure the momentum more accurately, we can't measure the position of the electron accurately.

Heisenberg realized this was vital in the interpretation of quantum theory. Indeed, he found that position and momentum were related to Planck's constant h. The relationship he arrived at can be stated simply as: the uncertainty in position (call it x) times the uncertainty in momentum (p) has to be equal or greater than $h/2\pi$. The involvement of h told us why we don't have uncertainties when we deal with large masses and distance (compared to those of atomic physics). Planck's constant is extremely small, and as masses and distances get larger the relationship has little significance.

Heisenberg's discovery became known as the *Heisenberg uncertainty relation*.[2] It told us that in the atomic realm you could determine only one of the conjugate variables x and p accurately at a given time. The more accurately you determine position, the fuzzier momentum gets. Similarly, the more sharply you focus in on momentum, the fuzzier position gets. Moreover, if you knew the position of the electron *exactly*, the uncertainty in its momentum would be infinite. Similarly, if you knew the momentum exactly, then the uncertainty in position would be infinite. In other words, you would have no idea where the electron is. It was almost as if position and momentum were at two positions under the microscope. When you focused in on one of the two, say position, the other went out of focus, and vice versa.

It's interesting to speculate on what led Heisenberg to his uncertainty relation. There were obviously many influences. When he finally published his paper, he referred only to Einstein. He had talked to Einstein at length in May 1926 after he had lectured at Berlin on his matrix theory. Heisenberg walked back to Einstein's apartment with him after the lecture, and they talked about matrix theory and its implications. Einstein preferred Schrödinger's theory, but he was impressed with Heisenberg and asked him many questions about himself. This was only the second time the two men had met, and the first time they talked only for a few minutes. The discussion made a strong impression on Heisenberg and no doubt had an influence on his discovery. But his discussions with Born, Dirac, and Jordan also had to have an influence. Dirac was at Copenhagen at the same time as Heisenberg, and he and Jordan had just formulated their transformation theory that showed clearly how matrix and wave mechanics were related and how they could be transformed back and forth. His discussions with Pauli were also no doubt influential.

Heisenberg was anxious to convey his results to Bohr, but Bohr was to leave for a skiing vacation in Norway early the next morning. This gave him time to polish his ideas. He soon found that there was a similar relation between energy and time.[3]

Heisenberg wrote up his results in a long letter to Pauli and mailed it to him. Pauli had tremendous insight into things such as this, and Heisenberg valued his opinion. If there was anything wrong, he would notice it. Pauli was, in fact, well known as a severe and sometimes caustic critic. To Heisenberg's delight Pauli wrote him a long letter back. He was tremendously enthusiastic about the idea and considered it to be an important breakthrough. Heisenberg was pleased; he now had Pauli's letter as an additional weapon when he presented the idea to Bohr.

Bohr also used his holiday to clear his mind. He had spent a tremendous amount of time thinking about the meaning of quantum mechanics and the problem of the dual wave-particle representation. He needed a rest. But it wasn't long before he was thinking of the problems again. In his mind the major difficulty centered around the wave-particle duality. Why were there two aspects to every physical problem? Why were they so different? It didn't seem as if the electron could be both a particle and a wave. After all, waves and particles acted quite differently. Waves passed through one another; admittedly, they modified one another as they did, but there was no dramatic collision between them. Particles, on the other hand, collided and in the process their properties changed significantly.

Bohr had also been thinking about the relationship between theory and experiment. But his point of view was different from Heisenberg's. This was no doubt a result of the two men's different personalities and outlooks. Heisenberg was very reliant on mathematics and quick to use equations in his search for answers. Bohr was more philosophical and interested in the physics behind the phenomena. He wanted to understand it thoroughly in his mind before he wrote down any equations. And Bohr had done considerable thinking about the problem, so he was primed for more discussions with Heisenberg and others at the institute when he returned.

BOHR'S RETURN

Heisenberg told Bohr about his discovery immediately after Bohr got back. Bohr was impressed. He realized it was an important breakthrough and was related to the things he had been thinking about. But as he looked closer at it, he became increasingly dissatisfied. Heisenberg had considered only the particle aspects of the problem. There was no discussion whatsoever of the wave aspects in his relations. Bohr was sure that both aspects had to be considered. In fact, he was sure the particle-wave duality had to be the starting point for any such discussion. Furthermore, Heisenberg had said nothing about the size of the microscope lens—its aperture. Bohr was convinced that it was important.

Bohr began thinking about the wave aspect in relation to Heisenberg's microscope. To close in on the position of the electron, you had to make the aperture very small and narrow so that few rays would pass through it. But as you made it narrower, you wouldn't be able to determine the momentum accurately. If you enlarged the size of the aperture, you could determine the momentum more accurately, but at the expense of losing accuracy in relation to its position. Following this line of argument, Bohr realized that there was a direct connection to the wave aspect of the particle. Waves passing through a small aperture or narrow slit are known to diffract or bend. As you narrow the opening more and more, diffraction increases. Because of this, you wouldn't be able to determine the momentum accurately.

Bohr also realized that energy and momentum were associated with the particle aspects of the electron, whereas position and time were associated with its wave aspects. This meant that in the uncertainty relation between position and momentum there was one wave variable and one particle variable. The same occurred for the energy-time uncertainty relation. Bohr was sure this was an important link between the particle and wave pictures, but Heisenberg did not agree.

Heisenberg was ready to publish his results and was reluctant to make any changes. Bohr argued that it was not ready for publication. The problem had to be thought through much more carefully, and the wave aspects of the problem had to be included to make it complete. Heisenberg disliked waves and the continuity they implied, and he was reluctant to change. Intense arguments followed. At one point, during a very heated argument, Heisenberg actually broke down in tears. He admitted later that he said things to Bohr that he wished he hadn't. But he now knew how Schrödinger must have felt and why he was driven to the sickbed. Bohr was, indeed, relentless. Relations between the two men became strained for a while. Heisenberg held back his paper and eventually realized that Bohr was right. He rewrote it so that it included several of Bohr's arguments, and on March 22, 1927, he submitted it to *Zeitschrift für Physiks*. The title of the paper was "On the Perceptional Content of Quantum Theoretical Kinematics and Mechanics." It was twenty-seven pages long and considered by many to be an even greater contribution to atomic theory than his matrix theory. For the first time the true limitations of quantum mechanics were established.[4]

COMPLEMENTARITY

Bohr's problem with Heisenberg's uncertainty relations centered on the wave-particle dualism and the act of measuring or observing an atomic system. It was obvious to Bohr that when you made an observation you disturbed the system, so it was no longer the same immediately after the observation. To him this was a highly significant point, and it would have serious repercussions on our ideas of reality and causality.

Heisenberg's relations, along with Born's probability interpretation of Schrödinger's wave function, had strong implications for the concept of determinism. Determinism had been a central part of physics since the time of Isaac Newton and Pierre-Simon Laplace. According to them, our universe was deterministic. In other words, if you knew the position, momentum, energy, and so on of a particle in the universe at one time, you could determine it for all time. In theory, its entire history would be known. In practice, of course, detailed calculation of it would be extremely complicated, but this was beside the point. Once you knew everything about a particle, you could predict, in theory, its fate for all time. With the new developments within quantum theory, this would no longer be possible.

The atomic realm, according to the uncertainty principle and the probability interpretation, was now indeterminate. If you knew the position, of a particle very accurately at one time, you didn't have its momentum. In fact, in measuring its position, you disturbed it so you didn't know where it would be a short time later. The universe of atomic physics was indeterminate, and, as we will see, this was to have important and even weird implications.

It had been assumed since Newton's time and earlier that the universe was a real universe independent of any observer. Measurement of such things as position, momentum, and energy did not change anything and had no effect on the fate of a particle. But within quantum theory this was no longer true. The act of measurement disturbed the system. Furthermore, according to Bohr, the particle did not become real until you measured it. It did not exist until it was observed, so it couldn't be real. This was, of course, quite a radical departure from what most people believed at the time.

Bohr also attacked the problem of wave-particle duality head-on. Although these two concepts seemed to be mutually exclusive, they were both important. Which aspect the electron exhibited depended on the experiment that you selected. Some experiments showed that the electron was a particle, others showed it was a wave. Moreover, the same went for light. The photoelectric effect, for example, showed it was a particle, and diffraction and interference phenomena showed it was a wave. But no experiment showed both aspects at the same time for either photons or matter particles. This meant that in selecting a particular experiment or type of observation you were not only making the particle real by observing it, but you were deciding which aspect it had.

The two aspects of the particle-wave dualism were complementary—exclusive, but they complemented one another. A good analogy is given by Fred Alan Wolf in his book *Taking the Quantum Leap*.[5] He uses a cube such as the one shown in figure 9. It is easy to see that you can visualize this cube in two different ways, depending on how you look at it. The face of the cube can be toward you on the lower-left corner, or it can be toward you in the upper-right corner. If you look at it one way, the other doesn't exist—it's still there, but you don't see it. The particle and wave aspects are the same. In observing a phenomenon in one way you see the particle aspect, but the wave aspect is still there. It's just that you're not seeing it. Because the two aspects are complementary, Bohr referred to the principle as the *principle of complementarity*.

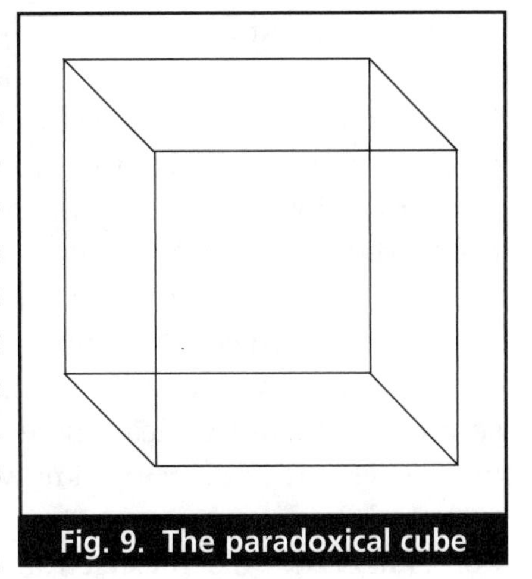
Fig. 9. The paradoxical cube

Inherent in the theory was another disturbing feature. It appeared as if the principle of causality was also being broken at the atomic level. Causality had been the foundation of classical physics. When something happened, it had to have a cause. For example, when a baseball was struck by a bat, it flew off into the outfield, and these two events had to come in the proper order. The baseball was thrown, the bat hit it, and it flew to the outfield. No one argued about this; it was accepted. It would have sounded a little crazy if you had tried to argue that the baseball flew out to the outfield before the bat hit it.

Strangely, we could no longer say this was true for events in the quantum realm. When an electron was hit by an energetic photon, if we knew the position of the electron very accurately, we would have no idea where it went, and we could not predict where it went. In the strictest sense, the present didn't determine the future, and therefore causality was violated.

Bohr's ideas along with Heisenberg's uncertainty principle and Born's probability interpretation of quantum theory eventually became known as the Copenhagen interpretation.[6]

THE WAVE PACKET

If the particle was a wave within Schrödinger's theory, it somehow had to be represented as such. And Schrödinger had represented it as a superposition of waves—a wave packet. Was this still valid? Many people did not like it, but they continued to use it. How would it be interpreted on the basis of the Copenhagen interpretation?

One of the major difficulties of Schrödinger's wave packet, as we saw earlier, was that it expanded.[7] In a very short time, it was so broad we had no idea where the electron was. The size of the wave packet represents our knowledge of the position of the electron, and its spread represents the uncertainty in momentum. We don't know how fast the electron is traveling because we have narrowed in on its position, so we don't know where it is going. If we make another observation of the electron, however, the wave packet suddenly shrinks and becomes sharp. After this measurement, it grows again, but a later observation will shrink it again.

What about the electron orbiting in the atom? Can we think of it as a wave packet? Indeed, we can, but we have the same difficulties as above. If we observe the wave packet in one orbit, we immediately knock it to a higher energy orbit. If we tried to observe it again, we might knock it out of the atom. Thus the word "orbit" is perhaps not a valid one in considering the energy states of an atom, but unfortunately we have nothing better.

THE COMO CONFERENCE

Although he had been intimately involved, Bohr had published almost nothing during the "miracle years" of quantum discovery, from July 1926 to September 1927, and he was anxious to publish his new ideas. His first opportunity to present them came at the Como Conference in Italy on September 16, 1927. It was in memorium of the hundredth anniversary of Allesandro Volta's death. Volta had made important discoveries in electricity and magnetism.[8]

Bohr's lecture was titled "The Quantum Postulate and the

Recent Developments of Atomic Theory." Many of the leading physicists in the world were present, but Einstein was absent. Bohr used the word "complementarity" for the first time in discussing what would become known as the *complementarity principle*. In the discussion that followed, no one objected, but it was such a strange idea it would have taken time for it to sink in, so most did not comprehend its full meaning.

Bohr had already started writing a paper, but for him writing a paper was a long, frustrating process with innumerable drafts. The paper was far from ready for publication. Fortunately, Pauli, who was at the conference, became tremendously enthusiastic about the new ideas and remained behind after the conference was over to help Bohr. Between the two of them, they soon had two papers ready for publication—one in German for *Naturwissenchaften*, and one in English for *Nature*.

THE SOLVAY CONFERENCE

About a month after the Como Conference came the Solvay Conference in Brussels. The theme was "electrons and photons." Bohr particularly looked forward to this conference because Einstein would now be in the audience. As it turned out, this was probably the most dramatic and important of all the Solvay Conferences. All of the major scientists of Europe were in attendance.[9]

Bohr expected Einstein to be swayed over to his point of view quickly. But Einstein was not easily convinced. He was disturbed by Bohr's ideas. He did not like the indeterminism of the new theory. He disliked Bohr's idea that there was no reality in the atomic world until something was observed, and he worried about the violation of the principle of causality. He was impressed with the successes of the new theory in solving previously unsolved problems, but deep down he felt this was not the last word. To him it was only an approximation, and beneath it was a better, more exact deterministic theory. But the Copenhagen interpretation forbade this.

One after the other, he presented ingenious thought experiments that seemed to show that Bohr's ideas were incomplete. But one by one Bohr overcame them. In each case he had to reach deep, but in each case he managed to refute Einstein. This went on day after day until the end of the conference. It was truly a "battle of the Titans." Bohr was insistent, but Einstein was not convinced. Each time Bohr overcame one of his arguments, Einstein would come up with another.

In the end Einstein had to concede that Bohr had done a good job. He admitted that from a logical point of view the theory along with Bohr's complementarity was consistent. Still, deep down, he couldn't believe it was the last word. Finally, in exasperation, his friend Paul Ehrenfest said to him, "Einstein, shame on you! You are beginning to sound like the critics of your own theories of relativity. Again and again your arguments have been refuted, but instead of applying your own rule that physics must be built on measurable relationships and not preconceived notions, you continue to invent arguments based on those same preconceptions."[10]

Bohr was disappointed that he didn't convince Einstein, but he took a certain pride in being able to show how each of Einstein's thought experiments were flawed. This, however, was not the end of the debate. And it would be many years before the Copenhagen interpretation was fully accepted. It took time for its significance to sink in.

HEISENBERG AND BOHR

Heisenberg felt bad after his sharp-tongued remarks to Bohr, and he wanted to make it up to him. But he wasn't sure how, and he was in an awkward position. He was still in Copenhagen at Bohr's institute, and he wanted to stay, but he was beginning to get offers from universities throughout Europe. If he accepted one of them and left, it might leave bad feelings between him and Bohr, and he didn't want this.

One of the most promising offers was from Leipzig University.

They were offering him a full professorship. If he accepted it, he would be the youngest full professor in Europe at twenty-six. He was tempted, but he turned it down. The same offer was made to Pauli, and he also turned it down. Sommerfeld then informed Bohr that he would be receiving money to expand his Institute of Theoretical Physics at the University of Munich, and a position would be available. Was he interested? Heisenberg was, indeed, interested; this was his hometown, and he was eager to return to it. But the position would not be available for a year or so. Sommerfeld offered to hold it for him if he wished.

University of Leipzig officials came after him again with an even better offer. Heisenberg talked to Bohr. Bohr was, of course, hoping to keep him in Copenhagen for a while longer, but he finally reluctantly advised him to accept the Leipzig offer. Interestingly, in the meantime he had two more offers, one from ETH in Switzerland and one from a university in the United States.

In October 1927 Heisenberg finally accepted the Leipzig offer. Years later he wrote, "Late in the autumn of 1927 I had to leave Copenhagen. . . . I returned . . . almost every year for a few weeks to talk over with Bohr the problems which occupied us both; but the period of close collaboration, that had been full to the brim with exciting scientific advances, and where I learned so infinitely much from Bohr, was unfortunately over."[11]

Chapter Nine

Einstein's Objections and Quantum Weirdness

It might seem strange that Einstein struggled so hard to show that quantum mechanics had shortcomings when he had contributed so much to the original theory. The determinism of classical mechanics, however, had been instilled in him from years of working with it, and he found it difficult to accept a world based on statistics and probability. It was too radical for him, even though, ironically, only a few years earlier he had been the radical, with views that seemed so weird and strange that few people could comprehend them.

Einstein was, nevertheless, impressed with the young radicals that formulated quantum theory, perhaps because they reminded him of himself when he was young. Despite his inner conviction that something was lacking in quantum mechanics, he was convinced that Heisenberg and Schrödinger deserved the Nobel Prize, and he nominated both of them in 1931. He definitely appreciated the importance of their work.

Einstein had presented several ingenious thought experi-

Albert Einstein

ments at the Solvay Conference in 1927, but to his chagrin Bohr had overcome them all. Nevertheless, Bohr hadn't convinced him, and when the Solvay Conference of 1930 approached, he was ready with a new paradox, and this one would take Bohr by surprise. He had no immediate response.

His thought experiment was as follows. Suppose you have a box that is filled with radiation, and on the box is a shutter that can release radiation. Furthermore, the shutter is attached to a clock, so whenever it is opened the time can be recorded. If you weigh the box before and after the shutter is opened, you can determine how much radiation was taken in or released, and since the shutter is connected to a clock, the exact time is known. Thus you know the energy within the box and the exact time. But according to the uncertainty principle, you cannot measure energy and time simultaneously.[1]

Bohr was caught off guard. He spent the evening discussing the problem with anyone who would listen to him. At last he conceded that he could see no way around it and decided to sleep on it. He was sure that once his mind was clearer he would be able to come up with something, and the next morning, to his delight, he came up with the answer. It was an embarrassment to Einstein, for Bohr used Einstein's own theory—general relativity—to show that there would be a slight change in the elevation of the clock during the weighing process. To be weighed the box would have to be suspended by a spring in a gravitational field, and when the photon left, the box would move slightly because it weighed less. If its position changed, there was obviously some uncertainty in its position in the gravitational field, and therefore some uncertainty in the rate at which the clock ran. And this uncertainty was exactly what was needed to bring back the uncertainty relationship between energy and time.

Einstein must have known that he was defeated, for he no longer challenged the foundations of the theory. Nevertheless, he was far from convinced and would continue challenging the theory in other ways for many years.

THE DOUBLE-SLIT EXPERIMENT

One of the major difficulties of quantum mechanics was the wave-particle dualism. It is brought out best in the double-slit experiment, and it shows that Bohr's concept of complementarity is, indeed, needed. Assume we have a single slit, a double slit, and a source of light. For the present we will assume that light is composed of particles, namely photons. If we project the beam of light toward the single slit, we see an image of the slit on a screen behind it, and this is what we would expect (fig. 10). With our assumption that light is composed of particles, it is natural. When we project the same beam on the double slit, however, we do not see an image of two slits on the screen. We see a series of lines, in other words, many images of the slit (fig. 11). We now refer to this as an interference pattern and know that it is caused by the interaction of waves. But how could it be explained in terms of particles?

To get a better idea of what is going on, let's project one photon at a time toward the slit and watch the image as it builds up on the screen. In the case of the single slit, after a considerable amount of time we see the same image we saw before—a single slit. Incidentally, there is a phenomenon called *diffraction* that causes faint lines near the edge of the slit, but for now we will neglect it since it has nothing to do with our argument. Turning now to the second slit, again we project photons toward it one at a time. As the image builds up, we see that we are getting the same multiple pattern we got before. In other words, there is a series of bright and dark lines

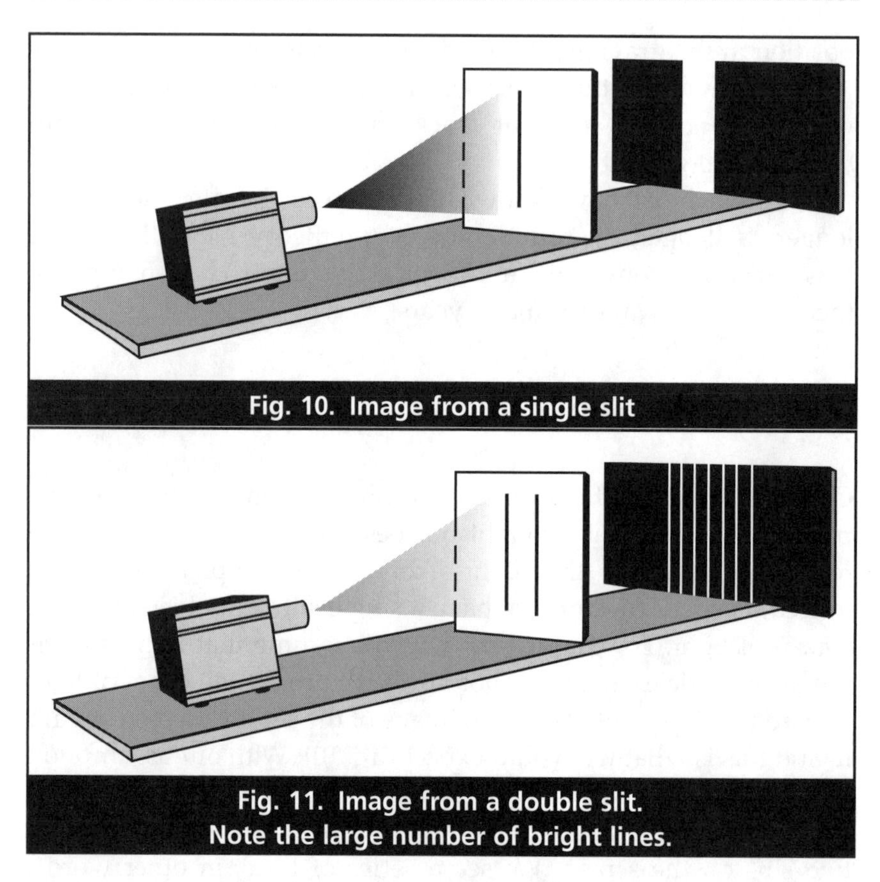

Fig. 10. Image from a single slit

Fig. 11. Image from a double slit. Note the large number of bright lines.

on the screen. If you think about this, you realize it is strange. In the case of a single slit, the image was a replica of the slit, and in the case of a double slit, the photons have to pass through one or the other of the two slits. A single photon, in passing through one of the slits, should have no idea that there is a second slit a short distance away. You would think it would go to the same point on the screen as a photon passing through a single slit. To go to a different point, it would have to go over to the right (or left) and check to see if there was a slit there. If there was a slit, it would go to one point on the screen, and if there was no slit, it would go to a different point. But a photon is a point particle. How could it move out as it approached the slit and check for a second slit? The only way it could do this is if it were a wave. Waves can spread out, and this is, of course, what we have to assume.[2]

For the moment, though, let's continue assuming that we are dealing with particles, and we've decided we would like to trick the system. We'll set up a small detector near one of the slits that can check to see what the photon did as it approached the slits. Did it somehow go over and check the second slit? But this creates a problem. According to the uncertainty principle, if we measure the position of a particle, we disturb it. It could be heading for one of the slits, and after we measure it, it could go through the other slit because of the disturbance. So we out-tricked ourselves; we're still not certain what happened.

EPR

After the 1930 Solvay Conference, Einstein knew it was hopeless to attack the foundations of quantum mechanics, but he still was not satisfied with the theory. He had little time to think about the problem, however, because of the turmoil in his life. The rise of Nazism in the early 1930s had worried him, and when Hitler seized power in January 1933, there was no question that his life was in danger. There had been considerable anti-Semitism earlier, but now it was much more out in the open and more overtly violent. Einstein, as one of the best-known Jews in Europe, was directly in the line of fire. Fortunately, about this time he had left Germany for a visit to the United States, and when he heard about the turn of events in Germany, he decided not to return.

Shortly after he left, Nazis broke into his house and confiscated most of his belongings. They seized his bank accounts, and he was attacked fiercely in the newspapers. Einstein counterattacked by renouncing his German citizenship and issuing statements against the new regime.

He didn't have to worry about a job in the United States. He had several offers and finally accepted a position at the newly forming Institute for Advanced Studies at Princeton, New Jersey. Life was now a little calmer, and he was able to get some work done. Within a couple of years, he published a paper that would have repercussions for years to come. It was directed at the completeness of quantum mechanics. Bohr had no doubt thought the arguments were over, and he had won, even though Einstein had not been convinced. But this paper showed that they were far from over. Einstein wrote the paper with two collaborators: twenty-five-year-old Nathan Rosen, who had come to Princeton in 1934, and Boris Podolsky, who was seven years younger than Einstein and had come from Caltech. The paper was published in the May 15, 1935, issue of *Physical Review* under the title "Can Quantum Mechanical Description of Physical Reality Be Considered Complete?" It is usually referred to as EPR, after the initials of the three authors.[3]

As I mentioned earlier, Einstein had accepted the mathematical basics of quantum mechanics, but because of the indeterminacy of the theory, he was sure it had to be incomplete. In a letter about this time, he wrote, "There is no doubt that quantum mechanics has seized hold of a beautiful element of truth and that it will be a test stone for any future theoretical basis, in that it must be deducible as a limiting case." In another letter he wrote, "On 'quantum mechanics' I think that, with respect to ponderable matter, it contains roughly as much truth as the theory of light without quanta. It may be a correct theory of statistical laws, but an inadequate conception of . . . elementary processes."[4] And finally, in a letter to Max Born he wrote, "Quantum mechanics is certainly imposing. But an inner voice tells me that it is not yet the real thing."

The arguments in the EPR paper centered around two basic assumptions:

1. The quantum mechanical description of a system given by the state function is incomplete.
2. Observables represented in the theory cannot have simultaneous reality.

The paper begins by setting down a condition for physical reality. "If without in any way disturbing the system, we can predict with certainty . . . the value of a physical quantity, then there exists an element of physical reality corresponding to the physical quantity." A definition of a "complete" theory was also given as follows: "Each element of physical reality must have a correspondence in physical theory." These two statements seemed reasonable, and most people did not argue with them.[5]

The essence of the paper was contained in a thought experiment that goes as follows. Consider two particles that have just collided. Call them *A* and *B*. Assume that after the collision they fly apart, and within a short time they are thousands of miles apart. We assume the conservation of momentum is valid, so if we measure the momentum of particle *A*, then we know the momentum of *B* even if it is thousands of miles away. In effect, we have determined this property without measuring it, and because we know it, it has to be real. But according to the Copenhagen interpretation, a particle does not become real until it is measured. In effect, it does not exist, and therefore particle *B* can't be real.

Furthermore, if we chose to measure the position of particle *A* rather than its momentum, we would know the position of the distant particle (particle *B*), and it would also have to be real from this measurement. The measurement of momentum of *A* would disturb its position, but this has no effect on the position of particle B.

Another way we can look at all this is by using the spin of the particle. I will present this argument because we will refer to it later. This approach is due mainly to David Bohm of the University of London.[6] As Samuel Goudsmit and George Uhlenbeck showed in 1925, particles have a property called *spin*. It's much like the spin of a spinning top, but we have to be careful in pushing the analogy too far. In quantum mechanics spin has only two states. The spin axis can be oriented in only one of two directions, namely up or down, and we refer to them as the *spins-up state*, and the *spins-down state*. Let's assume, then, that we have a two-particle system which is in the lowest energy state; this means that one of the particles will have spins-up and the other will have spins-down. Assume further that it remains in the lowest state and that the two particles fly apart. Within a short time, they will be separated by a long distance. If we measure one of the particles and find it to be spins-up, the other has to be spins-down. Thus we have determined a property of a distant particle without measuring it directly, which again is contrary to the Copenhagen interpretation.

According to the EPR paper, there are only two possible explanations of this outcome. Either quantum mechanics is incomplete, and this is what Einstein believed, or there is some sort of instantaneous communication between the two particles. This instantaneous communication would, of course, violate the principle of local causality, which states you cannot have instantaneous communication between distant objects. According to Einstein, "take your pick," and he was sure which of the two most people would pick. Everyone knew that local causality could not be violated, so it appeared obvious that quantum mechanics was incomplete. Einstein was convinced that this was the case.

As expected the paper was a bombshell in the scientific world. Pauli and Heisenberg were annoyed that Einstein was up to his "old tricks" again. Pauli urged Heisenberg to reply to it immediately. But Bohr had already taken up the challenge. According to Leon Rosenfeld, who was with Bohr when the news arrived, "It came down on us as a bolt out of the blue. Its effect on Bohr was remarkable."[7] According to Rosenfeld they dropped everything as they began looking at what Bohr was sure had to be a misunderstanding. Bohr immediately began dictating a reply, but soon discovered that refuting it was going to be a more difficult task than he thought. He and his colleagues worked on the reply for days, then weeks. Finally, six weeks later it was ready. It was published on June 29, 1935, in *Nature*. A longer, more detailed paper was later published in *Physical Review*. In his reply Bohr focused on two aspects of EPR. First was Einstein's definition of reality and second was the statement "without any disturbance to the system." According to Bohr both were ambiguous. It was impossible, argued Bohr, to speak of "physical reality" without including the measuring process in this reality, and therefore Einstein's arguments were invalid.

Einstein was amused by the reaction to his paper. He said he was bombarded by letters pointing out that he was wrong, but each of them thought he was wrong for a different reason. He admitted Bohr's arguments were the most logical and the best. Nevertheless, he did not accept them. The alternative to completeness was a strange "telepathic" communication between the particles, and Einstein was convinced that this could not be the case. The only one who seemed to be happy with EPR was Schrödinger, who wrote Einstein a congratulatory letter.[8]

THE SCHRÖDINGER-EINSTEIN LETTERS AND SCHRÖDINGER'S CAT

When the EPR paper appeared, Schrödinger was at Oxford, England. On June 7, 1935, he wrote to Einstein, "I am very pleased that in the work that just appeared in *Physical Review* you have publicly caught the dogmatic quantum mechanics napping over things we used to discuss so much in Berlin."

In Einstein's reply to the letter, we find something particularly interesting. He said that because of his problems with English, Podolsky had written the paper and that he hadn't seen it until after it had been sent for publication. He said that he was concerned that the paper had missed a critical point. He then proceeded to use a different example—the selection of a ball which could be in either of two closed boxes—in his attempt to make things clearer to Schrödinger.

After several more letters, Schrödinger became so involved that he published a lengthy article in *Die Naturwissenchraft* on November 29, 1935. It was titled "The Current Situation in Quantum Mechanics." The most interesting of its fifteen sections was the fifth in which he presents a paradox, now known as *Schrödinger's cat paradox*. Over the years this paradox has generated almost as much interest as EPR.[9]

Before getting into the cat paradox, let's consider how a quantum system is represented. It is represented by a wave function which describes the particles of the system and how they change in time, but it gives only probabilities. When a measurement is made, the wave function "collapses," and values of position, momentum, and so on are determined. What is important, of course, is that it is the act of measurement that gives these properties. Before this act, according to quantum mechanics, they do not exist.

In the cat paradox, a cat is locked in a steel cage with a radioactive source and a detector that can detect radioactive particles. If a radioactive particle is detected by the detector, a poisonous gas is released that kills the cat. The probability of emission of a radioactive particle in one minute is 50 percent. Assuming the cage is some distance away, we turn on the radioactive source remotely and wait for one minute.

Is the cat dead or alive at the end of one minute? According to the Copenhagen interpretation, it is neither. In fact, until we measure it, or observe it, it is neither dead or alive. This system is described by a wave function, and until we collapse the wave function, the cat does not acquire a definite state. In other words, it is neither dead nor alive. Most people find this interpretation hard to accept.

In this example Schrödinger is, of course, pushing the predictions of quantum mechanics in the microworld to the macroworld, and they don't necessarily apply here. Nevertheless, it's an interesting example.

BELL'S INEQUALITY

If quantum mechanics is indeed incomplete, as Einstein asserted, then it must have what are called "hidden variables." This would mean there is a subtheory, which is presently beyond our insensitive measuring techniques. Although it is presently hidden, one day we will presumably detect it. The idea of hidden variables has been around for many years. Several hidden-variable theories have been put forward; two of the better-known ones are those of David Bohm and Louis de Broglie.

Einstein supported the idea of hidden variables, but he was not a strong advocate, and he never tried to formulate a hidden-variable theory. His view was that a completely new theory was needed, and he spent many years searching for it in his quest for a unified field theory. This was an extension of his theory of general relativity—a complete theory that would include quantum mechanics as a special case, but within the theory it would not be indeterminant.

The idea of hidden variables suffered a setback when John von Neumann of the Institute for Advanced Study in Princeton presented a proof that hidden variables could not exist. This proof was included in the fourth chapter of his book *The Mathematical Foundations of Quantum Mechanics.* In this chapter von Neumann considers whether quantum mechanics is a logically complete theory, or "whether it could be formulated as a deterministic theory by the introduction of hidden parameters, that is, additional variables, which unlike ordinary variables, are inaccessible to measurement and hence not subject to the restrictions of the uncertainty principle." He concludes, "The present system of quantum mechanics would have to be objectively false in order that another description of the elementary process than the statistical one may be possible."

This seemed to settle the issue. Papers were still published periodically, but there was not a lot of interest. Then in 1964 John Bell of CERN, who was on leave from the University of Wisconsin, published a paper in *Physics.* The following year he published a longer paper in the *Reviews of Modern Physics.* In the first of these papers, he showed that von Neumann's proof that hidden variables were not possible was too stringent. He went on to construct a hidden-variable theory using spinning particles. In the second paper, he introduced his famous *Bell Inequality.* In it he showed that any "local" hidden-variable theory cannot reproduce all the statistical predictions of quantum mechanics. By local we meant that it happens at a precise location. A local hidden-variable theory therefore affects things only at a precise location.[10]

Bell showed that local hidden variables would, indeed, produce results that contradicted the predictions of quantum mechanics. His proof was in the form of an inequality in the relationship between the spins of particles. We saw earlier that Bohm explained the EPR paradox in terms of spin, and Bell used this formulation in setting up his inequality.

Let's consider how we measure spin. In a macroscopic object such as a spinning top we measure all three components of spin in space (i.e., along the three directions of space, namely x, y, z) then use them to determine the overall spin. But in the quantum world, there are only two directions of spin: up and down. Suppose that we measure spin along one axis and find it is up, then we go to another axis and measure it, finding it is down, then finally we go back to the first axis and remeasure it. We will find that when we remeasure it, it will not be in an up state. In fact, it will be in the same state only 50 percent of the time. With the first measurement we have disturbed the system, so we're not going to get the same value when we remeasure.

So, what is the actual spin of the particle? We don't know for sure. Furthermore, if this particle is connected with a second particle as in our previous example, we have problems. Separately, the two particles seem to flip back and forth from one spin state to the other, but their total spin always has to be zero. How can this be? The only way, it seems, is that the two particles are connected by a sort of action-at-a-distance. In other words, when one particle flips to a particular state, it signals the other particle so that it can flip to the opposite state.

Bell's inequality showed that there was a relationship between the spins of two particles in a real experiment involving correlations such as those above. The critical thing was that the experiment was no longer a thought experiment; it was real and could be performed in the laboratory. We should be able to measure the spin along a particular axis for one particle and the spin along a different axis for the second particle. In the experimental setup, the spins of large numbers of pairs of protons (and photons) were measured. The results were checked against the Bell inequality to see if it was valid, or perhaps, violated.

The first tests were performed at the University of California, Berkeley using photons in 1972. Over the following years, several more tests were performed and by 1975 six tests had been completed. Of these, four violated Bell's inequality. Violation of the inequality favored quantum mechanics as a complete theory and

Bohr's Copenhagen interpretation. So it appeared that Bohr had won and Einstein had lost, but there was still some uncertainty.

Further experiments were carried out in the mid-1970s by other experimenters. The physical setup was different. Electron-positron annihilation provided gamma-ray photons that were correlated. In this case five out of seven tests favored Bohr and again it appeared as if he had won, but there was still a chance that there was a loophole. The most comprehensive and accurate tests were then performed by a team at the University of Paris under Alain Aspect in 1982.[11] Aspect used photons along with a highly accurate method of determining particle spin, and his results were decisive. Bell's inequality was definitely violated, and there appeared to be no way around the conclusion. Einstein's assertion that quantum mechanics was incomplete could not be true. There were no hidden variables in the theory.

This was a shock to many physicists. The reason was that the only alternative, according to the EPR paradox, was that local causality was violated. This meant there was some sort of strange "connectedness" between the two particles and it was instantaneous. But nothing can travel at a speed greater than the speed of light according to relativity. Yet the two systems seemed to be able to communicate with one another at a greater speed. Einstein referred to this mockingly as "telepathic" communication, and he was strongly against it. The only conclusion we can come to is that quantum mechanics is not incomplete and there are no hidden variables.

CONSEQUENCES

It may seem strange that so much attention has been paid to the philosophical aspects of quantum mechanics. After all it has nothing, or at least very little, to do with the formulae and numbers that come out of the quantum mechanical calculations. Quantum mechanics gives us a recipe, even if it does seem a little strange at times, for solving any atomic problem, and the results of experiments on Bell's inequality have no effect on what comes out of the calculation.

Why, then, do we spend so much time worrying about such things? One reason has to do with whether it is useful to look for another theory beyond quantum mechanics. Is quantum mechanics the end of the line? According to EPR and the tests of the Bell inequality, it is. There are no theories beyond it that will give us a better answer. In particular, there are no theories that will give us a deterministic answer.

The Copenhagen interpretation is still considered by many to be somewhat obscure, and not everyone is happy with it. We will see that quantum mechanics still had to be made relativistic. In other words, an equation that applied at very high velocities, close to those of light, was needed, and it soon came. There was also the problem of the interaction between particles and photons. It still had not been solved, and as we will see a new approach was needed for it.

Chapter Ten

Extending the Theory

Heisenberg's and Schrödinger's theories gave a beautiful explanation of most atomic phenomena, but there was one problem that wasn't solved adequately, and it was the interactions between photons and electrons, or more generally, the interactions between matter and fields. The idea of a field had been put forward many years earlier by Michael Faraday, and it had been put on a firm mathematical footing by James Clerk Maxwell. Maxwell formulated four equations that completely explained all of electricity and magnetism and the electromagnetic field.

A field can take many forms. From a simple point of view, it is a region where a quantity is specified at each point throughout the region. A simple example would be the wind velocity at all points in a given region. The two fields you are likely most familiar with are the magnetic field and the gravitational field. A magnetic field exists in the region around a bar magnet. We can specify the magnitude and direction of the field at each point near one end of the magnet. In the same way, we can specify the magnitude

Paul Dirac

of the gravitational field at each point above Earth's surface.

Once quantum theory was well established, scientists became interested in applying it to fields. Schrödinger's equation gave exact results for problems related to matter and matter particles but seemed to be limited when applied to fields. In 1927, however, Paul Dirac showed that both matter and fields could be quantized, and he published a theory showing how photons of light could be absorbed and emitted by atoms and electrons. It was a successful theory, but it was limited. It did not apply to high-speed particles, and in practice most reactions between matter particles and photons occurred at very high speeds.

What was needed was a relativistic theory, and again Dirac came to the rescue. In 1928 he published what would eventually become known as *Dirac's equation*. It was, in a sense, a relativistic version of Schrödinger's equation, but it was even more than this. It predicted the spin of the electron, and it allowed scientists to calculate the positions of spectral lines in hydrogen to an unprecedented degree of accuracy. But strangely, it predicted things that did not appear to exist. In particular, it predicted both positive and negative energy particles. What was the significance of negative energies? No one had ever seen a negative-energy particle. Did such things exist?

Dirac pondered the problem and soon came up with an answer. He visualized a "sea" of negative-energy particles that existed all around us, but this sea was filled and because of this we couldn't see it. It was invisible. He then went on to postulate that positive-energy particles could not make transitions to the negative energies because the negative-energy sea was filled, but there was the possibility of transitions from the negative sea to the positive energies.[1]

What would we see if such a transition took place? According to Dirac we would see a positive-energy electron and a hole in the sea of negative energies. The hole would appear as a particle of opposite charge. Since the electron had a negative charge, it would be positively charged. But no particle similar to the electron with a positive charge had ever been seen. Was it possible this particle was the proton? It didn't seem likely since the proton was much heavier than the electron. Furthermore, Robert Oppenheimer soon showed that the hydrogen atom wouldn't be stable if the new particle had a mass different from that of the electron.

The only alternative seemed to be a particle that had not yet been discovered, and Dirac predicted that this was the case. The

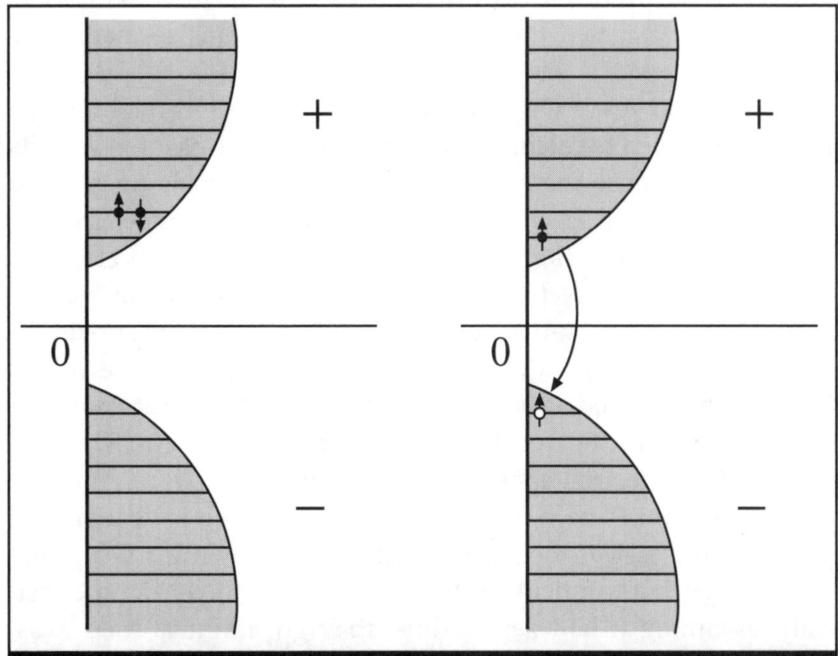

Fig. 12. The illustration to the left shows filled negative sea. The illustration to the right shows an electron jumping to positive sea leaving a hole.

new particle would be similar to the electron but have a positive charge. He referred to it as the *antielectron*, but many soon began referring to it as the *positron*. To everyone's surprise the new particle was discovered by Carl Anderson a couple of years later. But what about other particles such as the proton? Did they have oppositely charged antiparticles? According to Dirac they did, but it was twenty-five years before the first of these—the *antiproton*—was discovered.

Dirac's sea of negative-energy particles was a useful way of looking at the problem, and it was helpful in many respects. Today, however, we no longer view the antiparticle in this way. Nevertheless, Dirac's equation is still one of the most important equations in all of science. It is the basis of what we now call *quantum electrodynamics*—the theory of the interaction between fields and matter particles, or more exactly, the interaction between electrons, protons, and photons. This theory has allowed physicists to solve problems that were previously unsolvable. It seemed that physicists had, at last, tied the final knot. All problems seemed to be solvable.

In 1930, however, Oppenheimer showed that this might not be the case. He tried to calculate the interaction of the electron with its own field. Because the electron is a charged particle (negatively charged), it has an electric field around it. Oppenheimer tried to calculate the interaction energy associated with this field; it is referred to as the electron's *self energy*. (Self energy can also be thought of as the back-reaction of a particle on itself.) Surprisingly, he found he couldn't do it, or at least the answer came out to infinity. This seemed a little crazy. It had to be finite. But Pauli and Heisenberg soon showed that Oppenheimer was right. The self-energy of the electron appeared to be infinite.

To some this was not a surprise. After all, we have a similar problem in classical theory. Physicists had known for years that when they tried to calculate the self energy of the electron classically, assuming it had zero radius, they got infinity. There was, however, a way around this: give the electron a tiny, but finite radius. This seemed to be the way to proceed in quantum mechanics, but when scientists tried it, it didn't work. The self energy was still infinite.

The problem of self energy appeared to be associated with a cloud of photons that surrounded the electron. Such a cloud had been predicted several years earlier. To understand how such a cloud could exist, we have to go back to the uncertainty principle. It predicts a "fuzziness" associated with time and energy at the atomic level. Because of this fuzziness, an electron can emit a photon as long as it reabsorbs it within a very short period of time. This time depends on the energy of the photon. In essence, the photon sneaks out under the cloak of the uncertainty principle, but it has to get back under the same cloak. Strangely, it doesn't have to get back to the same electron; it can go to a different one, and, as we will see later, it frequently does. Such photons are called *virtual photons*. We will never be able to see them directly, but we have ample evidence that they exist. Indeed, there is a whole cloud of these virtual photons surrounding the electron.[2]

The self energy problem was the first, but it wasn't the last. Within a short time another problem arose, and it was associated with the mass of the electron. It was well known that part of this mass was due to its surrounding field. But as we closed in on the electron, this field increased dramatically; in fact, very close to the surface of the electron it approached infinity. What this meant was that the overall mass of the electron was infinite—at least according to the theory. It was well known experimentally, however, that this wasn't the case. The electron had a finite mass.

Physicists had barely recovered from the shock of this when it was discovered that the charge of the electron also had to be infinite. Again, this was due to the photon charge cloud. Photons don't have a charge, so they were not a problem, but photons very close to the electron had exceedingly high energies, and these energies were high enough for electron pairs to be generated, and they were charged. These pairs lasted for only a brief instant, but this was long enough to have an effect. The electron of the pair, with its negative charge, was repelled by the electron, but the positron was attracted and moved slightly closer to the electron. The result was a "screening" of the true charge of the electron. With this screening the electron could have an infinite charge.

In short, the "bare" or unscreened charge of the electron was infinite, but it was screened down to the charge we see by the virtual pairs that were being created around it. This screening is now referred to as *vacuum polarization*. By 1936 theorists had become convinced that the major problem with infinities in quantum electrodynamics was related to vacuum polarization and the infinite mass and charge of the electron. If progress was to be made, a method had to be found that would take them into account. Victor Weisskopf of MIT suggested that it might be possible to redefine mass and charge, a process we now refer to as *renormalization*. He pointed out that it would have to be done carefully, and it would be a complex problem.

But not everyone took his suggestion seriously. After all, so far no one had ever measured the small errors or corrections that these effects would create. They were second-order effects, and beyond our instrumentation. Or were they? As early as 1933 a number of experimentalists had begun to find slight discrepancies with predictions, but there was considerable controversy over this. William Houston of Caltech and Y. M. Hseih of China measured the fine structure of hydrogen and found what they believed was a discrepancy when they compared it with predicted values. W. E. Williams of London found similar effects, but they were so small not everyone believed they were significant. It was still possible that it was an experimental error. Bohr and Oppenheimer, however, were not convinced. They were sure that vacuum polarization had to be taken into account. Then the war came, and for several years little was accomplished.

During the war a number of people developed skills in radar and microwave techniques. Among them was Willis Lamb. Lamb had completed his Ph.D. under Oppenheimer in 1938. His interest in microwave spectroscopy developed at Columbia University where he worked with I. I. Rabi. Several new devices and techniques were developed as a result of the war effort, and Lamb soon incorporated them into his work. Shortly after the war ended, he became interested in the fine-structure spectra of hydrogen. After reading about the work of Houston and Williams, he began looking into whether the new techniques that had been developed during the war would allow him to do the experiment more accurately. He was soon convinced that they would, and he began building the equipment needed to perform the experiment.[3]

Willis Lamb

According to the theory Dirac had presented, there were two states in the hydrogen atom with slightly different energies. So far, though, no one had been able to distinguish them. They appeared to have the same energy. Lamb reasoned that if microwaves were passed through hydrogen, the energy absorbed by the two states would be slightly different, and this difference would be measurable. His first attempts were unsuccessful, but he kept at it. Then he began working with a graduate student, Robert Retherford, who had a detailed knowledge of the techniques and devices needed in the experiment. The apparatus was redesigned, and finally they were able to measure the energy difference of the two states. In fact, they were able to measure it incredibly accurately to many decimal places. It is now referred to as the *Lamb shift*. They were also able to measure what is called the fine-structure constant; it is a combination of the electronic charge, Planck's constant, and the speed of light. For years it had played an important role in physics and was known to be approximately 1/137. Lamb showed its exact value was 1/137.0365.

Lamb was awarded the Nobel Prize for his work. He was informed he had won the prize while he was teaching a class. He took the news calmly, then continued teaching his class. Not until it was over did he talk to reporters.

The Lamb shift was a second-order effect in quantum electrodynamics, and now that it had been accurately measured, theorists realized it was time for changes in the theory. Second-order effects such as self energy appeared to be infinite according to the calculations, but now it was evident that they were finite even if they were small. A conference was called to address the problem. It took place in June 1947 at Shelter Island, a small island off Long Island, New York. By today's standards it was small, but most of the major quantum theorists were there, including Robert Oppenheimer, Hans Bethe, Victor Weisskopf, Hendrik Kramers, Julian Schwinger, and Richard Feynman. Lamb was, of course, invited to report on his findings. With the tremendous interest in his breakthrough, he was the star of the conference.

The central concern of the group was the renormalization of quantum electrodynamics. The infinities that occurred in the second-order effects would have to be eliminated, but they would have to be dealt with properly. The meeting opened on June 2 with Willis Lamb reporting the details of his experiments. Discussions then centered on how to attack the problem of the deficiencies of the theory. Renormalization, or "redefining the constants of the theory," seemed to be the obvious solution, but everyone knew it was going to be difficult.

Hans Bethe thought about the calculation while he was on the train back to Cornell after the conference. He knew the full-blown relativistic calculation would be exceedingly complex, but the nonrelativistic one should be manageable. In fact, it would be a useful guide and would show if the idea of renormalization would work. He began the calculation on the train and finished it a few days later at Cornell. To his surprise it was easier than anticipated. Furthermore, it was a success. Even though it was a crude calculation, it accounted for 90 percent of the discrepancy. Bethe was sure that the additional 10 percent would be accounted for when the full relativistic calculation was performed. News of Bethe's success soon spread, and his quick success surprised many of the participants.[4]

THE BREAKTHROUGH

Several people returned from the conference eager to begin work on the problem. Among them were Julian Schwinger, Richard Feynman, and Willis Lamb. Even though Lamb had worked for several years as an experimentalist, he had been trained as a theorist under Oppenheimer. He and a student, N. Kroll, were the first to solve the problem, but their method was clumsy and unreliable. Schwinger was also anxious to try his hand at the problem, and with his tremendous talent it wasn't long before he had solved it.

Long, detailed calculations came easily to Schwinger. He had been known to make calculations of hundreds of pages without an error. A child prodigy, he was so brilliant in school he was able to skip most of high school. He went directly to Columbia University, completing his bachelor's degree in less than two years, and by the time he was twenty-one Schwinger had his Ph.D. In addition, he had already published several scientific papers by this time.

Not only did Schwinger solve the problem of renormalization, but he calculated the magnitude of another second-order effect related to the magnetic field of the electron—the *anomalous magnetic moment* (a small correction to the magnetic moment, or the extent to which the system looks like a magnet). It was a horrendously complicated calculation, and he did it in record time. In his attack on the problem of renormalization and the Lamb shift, he literally started at page one. He began by carefully reconstructing all of quantum electrodynamics and showing how each of the

Julian Schwinger

difficulties could be overcome. It was a long and arduous task involving hundreds of pages of calculations, but in the end he tied everything together and renormalized the theory, accounting exactly for the discrepancies.

Meanwhile Richard Feynman of Cornell also began working on the problem. Like Schwinger he had been born in New York City. He received his bachelor's degree from MIT in 1939 and Ph.D. from Princeton University in 1942. And like Schwinger he was a genius of the first order. He astounded people with his ability to perform complex calculations in his head. Indeed, he liked to bet people that he could solve any problem in sixty seconds that they could state in ten seconds, and he usually won. He was also well known for his antics; he liked to play the bongo drums late at night, and while working on the atomic bomb at Los Alamos, he developed the ability to crack almost any safe. As a joke he would break into a "foolproof" safe and "borrow" top secret documents, leaving a note that he had left them in the top drawer of a nearby desk. Most of the top officials did not find these jokes funny.[5]

It's interesting that just before he made his breakthrough in quantum electrodynamics, which was his greatest contribution to physics, he was getting disgusted with physics. He felt burnt out and was sure his days of creativity were over. He was also depressed because his wife of a few years had just died of tuberculosis. He had little interest in tackling formidable problems such as the renormalization of quantum electrodynamics. Then one day he was in the cafeteria when one of the waiters began tossing dishes in the air. He watched as the dishes wobbled strangely. Why did they wobble this way? Then it struck him: this was the type of problem he loved to solve. For him it was a challenge, and most important, it was fun. Over the past few years, most of the fun had gone out of physics for him. He went to Bethe's office and talked to him about renormalization. Bethe had begun to work on the relativistic problem but had got bogged down. "I know what the problem is," said Feynman. "I'll have the calculation for you in a couple of days." Bethe wasn't sure whether to take him seriously or not, but he knew he was capable of it.

Feynman went back to his office and started working. Within a short time he had everything worked out; the infinities were gone, and the results were finite and agreed with experiment.

The following day he was back in Bethe's office. The two men went to the blackboard, and Feynman went through his calculation. To his annoyance it didn't work out. Something was wrong. He went through it again carefully, but it still didn't work. The infinities were still there. In a state of confusion, he went back to his office and went carefully though everything, and this time it worked. The infinities were gone. Relieved, he headed back to Bethe's office and went through the calculation again, and this time everything came out as expected. The infinities were gone, and the theory was renormalized. Feynman later said that he never did figure out what had gone wrong the first time around.

With the success of the Shelter Island conference, a second conference was called the following year. It was to be held in Pennsylvania. Again, all the major theorists in quantum theory were in attendance. Neils Bohr was also there this time. The first person to speak was Schwinger. He outlined his method, and it was soon obvious to everyone in attendance that he had performed a tremendous feat. Everyone, including Bohr, was in awe as he went through the details. There was no doubt: the problem had been solved. But not everyone was happy. The mathematics that he used with such agility was extremely complex. It almost seemed as if no one but a Schwinger would be able to solve problems using his technique. Nevertheless, most people were impressed.

Then came Feynman. His approach was entirely different. Rather than fill the blackboard with complex equations, he drew funny little diagrams and used them to form his equations. To many his technique was confusing; they were not quite sure what he was doing. Bohr, in particular, was perturbed and kept questioning Feynman about each step he was taking. Feynman had developed his own approach to quantum mechanics and used it freely. Bohr didn't understand it and in frustration finally told him he should go back and learn traditional quantum mechanics. Feynman was not pleased with the insult and left the conference in

Richard Feynman

a gloomy mood. Back at Cornell he convinced himself that his procedure was easier and better than Schwinger's, and he would prove it. He wrote up several papers and sent them to *Physical Review*. And sure enough it was Feynman's method that caught on. Within a few years almost everyone was using it. Feynman joked that within a few years *Physical Review* would be full of his "funny little diagrams." And his prediction came true.[6]

Let's look at Feynman's method in a little more detail. He developed his diagrams so he could keep track of all the mathematical terms that went into his final expressions. Consider the collision of two electrons. The two electrons would repel one another, and they would do this by passing photons back and

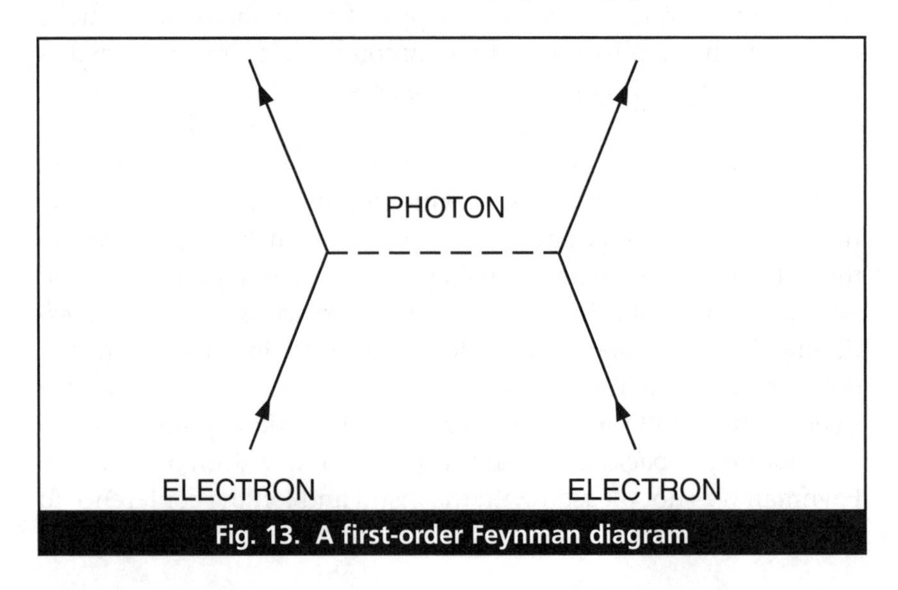

Fig. 13. A first-order Feynman diagram

forth. These photons would be absorbed and emitted by the two electrons. He represented the paths of the two electrons by solid lines, and the virtual photons by dotted (or squiggle) lines (see fig. 13). The point where the dashed and solid lines joined was called a *vertex*. This is where photons are absorbed or emitted. Mathematical terms were associated with incoming lines, vertices, and outgoing lines. Using these diagrams, Feynman was able to write down his expressions in terms of the contributions from them.

Feynman showed that he could represent any reaction between electrons, protons, and photons using his diagrams. Diagrams such as figure 13 were first order, but renormalization was a second-order effect. Feynman showed, however, that similar diagrams could also be used for second-order effects. As we saw earlier, the generation of an electron-positron pair from a virtual photon could occur if the energy was high enough, and close to the electron if it was high enough. Feynman represented this as follows.

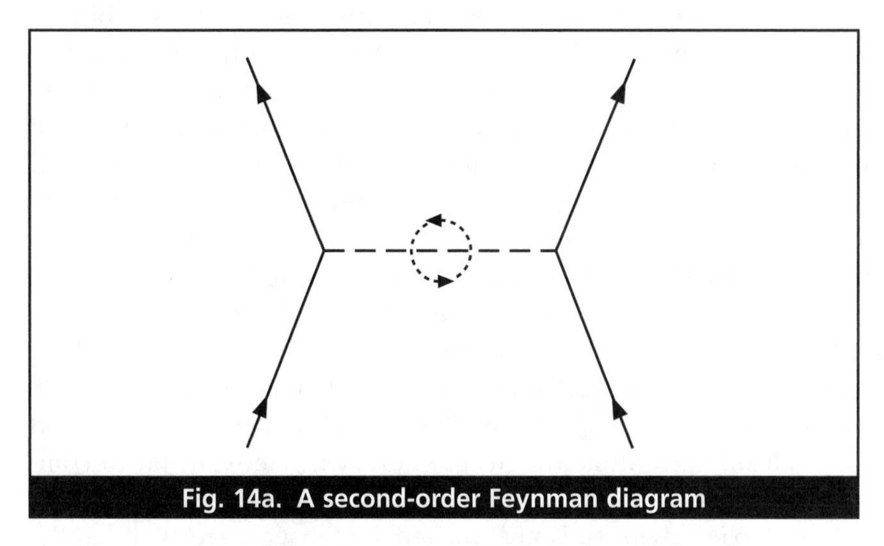

Fig. 14a. A second-order Feynman diagram

It is a second-order diagram, and as with first-order diagrams each line and vertex corresponds to a mathematical term which contributes to the overall expression. There were many other second-order diagrams that had to be included. Another one is shown below.

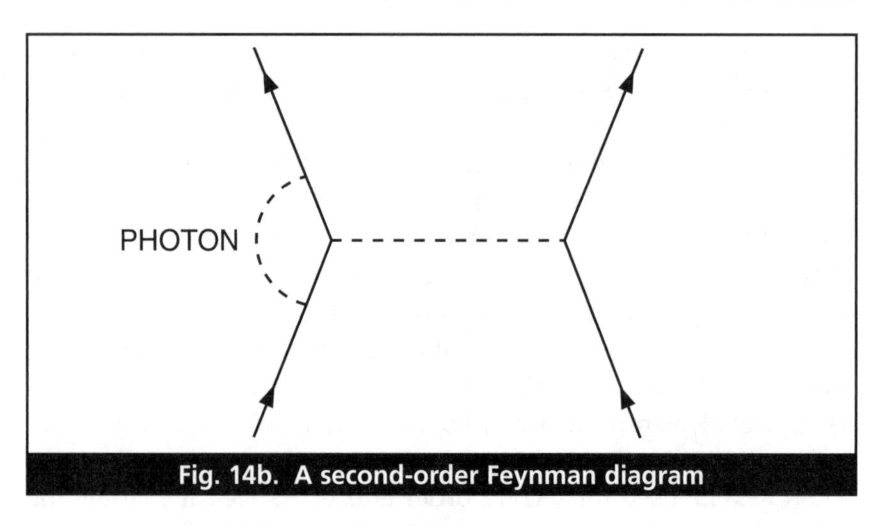

Fig. 14b. A second-order Feynman diagram

Two methods for renormalization were therefore presented at the Pennsylvania conference. But surprisingly a third method surfaced at about the same time. When Oppenheimer got back to his office after the conference, a manuscript was waiting for him. It contained a third, and apparently different, solution to the problem. Shin'ichiro Tomanaga of Japan had derived the technique in 1943, several years earlier, but because of the war it had not been published, and almost no one knew about it. Tomanaga had worked with Heisenberg in Germany before the war and was quite familiar with the problems of quantum electrodynamics.

It almost seemed that history was repeating itself. Earlier there had been two apparently different formulations of quantum mechanics (or three if you include Dirac's version), and now there were three different formulations of renormalizable quantum electrodynamics. In 1947, however, Freeman Dyson of Cornell showed that all three formulations were equivalent; he went on to complete a proof of renormalization to all orders of approximation.

In 1965 Schwinger, Feynman, and Tomanaga were awarded the Nobel Prize for their work. Strangely and tragically it was later discovered that another solution had been obtained as early as 1942. Ernst Steuckelberg of Switzerland solved the problem and submitted it to *Physical Review*. The referees did not understand what he had done and rejected it. In the meantime Steuckelberg's health was rapidly deteriorating. He struggled to revise the paper, finishing in 1945, but he never published it in a journal. It only appeared in the thesis of one of his students, and by that time Schwinger, Feynman, and Tomonaga had already presented their results.

With the renormalization of quantum electrodynamics, it seemed as if physics was complete. In theory, any known problem could be solved using the techniques that had been developed. But it was soon discovered that electromagnetic interactions were not the only ones that occur in nature. Electromagnetic interactions in the atom hold the electrons in place, and they're also responsible for the interactions between electrons and protons. But there is a much stronger force holding the particles of the nucleus together. It is referred to as the *strong interactions*.[7] Theorists assumed that the same techniques that work so well for quantum electrodynamics would work for the strong interactions. But to their surprise, they didn't. Second-order effects, which in quantum electrodynamics are 1/137 times smaller than first-order effects, were larger than first-order effects in the theory of strong interactions. This didn't make sense. How could second-order effects be larger than first-order effects?

Then another type of interaction was discovered, called *weak interactions*. And again renormalization was tried, and again it gave numbers that were greater than first-order effects. Something was obviously wrong. But that's another story.[8]

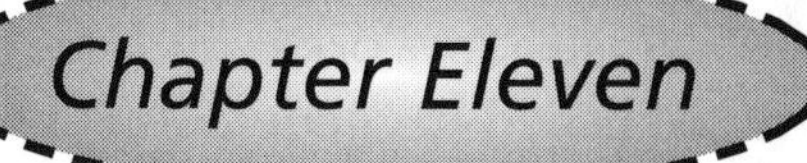

Modern Developments

Lasers and Masers

Over the next few chapters, I'll be discussing the numerous "spin-offs" of quantum mechanics. It may surprise you, but they cover many of the modern devices you use every day, and I don't think it's an exaggeration to say that the modern world would be quite different if quantum mechanics had never been discovered.

I'll begin with *lasers* and *masers*. They are now all around us; you see a laser in use every time you go to the grocery store. It's amazing, in fact, how prevalent they have become in modern society. Still, a laser beam is nothing more than an ordinary beam of light. Actually, it's not quite as "ordinary" as the light that is emitted from a light bulb, but it's light nevertheless. A laser beam does have distinct properties, as we will soon see.

EINSTEIN AND THE LASER

The origin of the laser can be traced to a paper Einstein published in 1916. If you're familiar with Einstein's work, you'll recognize 1916 as the year in which he published his general theory of relativity. It was, indeed, a busy time for him; in all, he published fourteen papers that year. Most scientists would be happy if they managed to publish a couple of papers in a year. And not only were Einstein's papers important, they were papers that literally changed the world. Besides his general theory of relativity, he published the first paper on *gravitational waves*—waves that are created by oscillating matter—in addition to three landmark papers on quantum theory. Furthermore, he completed his book *On the Special and General Theory of Relativity*, which is generally considered to be one of his best books.[1]

The first of Einstein's three papers on quantum theory contained an elegant, new, and extremely simple derivation of Planck's law of radiation. In this paper he introduced the concept of a *transition probability*. Bohr had introduced the idea of energy levels several years earlier and had speculated that electrons would jump from one energy level to the other. Furthermore, he said that when they jumped to levels of less energy, they would emit photons; and when they absorbed a photon, they would jump to a more energetic level. But Bohr said little about how this transition took place. It was Einstein who explained it.

Einstein discussed two types of emission in this landmark paper. The first type occurs when an electron jumps down and emits a photon. It is referred to as *spontaneous emission*, and it occurs when an electron is in an excited state. The electron can stay in this excited state for only a very short period of time. After a millionth of a second or so, depending on the conditions and type of atom, it will jump downward to a lower energy level, usually its ground state.

Einstein showed that there was also another type of emission, which we now refer to as *stimulated emission*, or *induced emission*. It takes place when a photon strikes an electron that is already in an excited state. Instead of absorbing the photon, the electron is stimulated to fall back to the ground state. When it does this, it emits another photon. So in this case there are two photons associated with the emission, and, of particular importance, they have exactly the same properties. In other words, they have the same wavelength and their waves are in phase.[2]

Einstein's paper was an important one, and it soon became the basis for many significant developments. But, strangely, the importance of stimulated emission was not realized for many years. It was discussed occasionally in theoretical papers, but its potential was not realized. Richard Tolman of the University of California mentioned it in a paper in the early 1920s. In his paper, which was titled *Duration of Molecules in Upper Quantum States,* he wrote, "Molecules in the upper quantum states may return to the lower quantum state in such a way to reinforce the primary beam by negative absorption."[3] The negative absorption he is referring to is, of course, stimulated emission. The statement also implied that a light "amplifier," or device that would amplify light, was possible.

For the next ten years, there was little interest in the phenomenon of stimulated emission. As we saw earlier, Lamb and Retherford used the methods of radiofrequency spectroscopy to measure the difference between two second-order spectral lines in hydrogen. Lamb admitted later, however, that they were so concentrated on their project that they didn't think of examining the properties of *negative absorption* or stimulated emission in microwaves, even though they used microwaves in their work. But progress was soon made.

THE MASER PRINCIPLE

Radar was one of the major discoveries of World War II, and because of its value in detecting planes, considerable research went into it. After the war many former military personnel with expertise in the area of microwaves continued to pursue their interest in the field. Considerable attention was directed at the interaction between matter particles and microwaves, and a new branch of research known as *microwave spectroscopy* was soon developed.

Some of the first work on the theory of microwave amplifica-

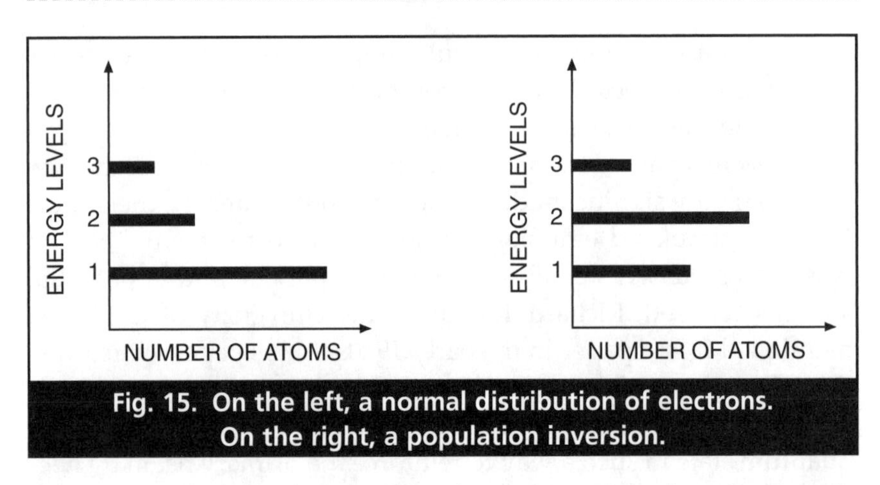

Fig. 15. On the left, a normal distribution of electrons. On the right, a population inversion.

tion using stimulated emission of radiation was done by Joseph Weber.[4] Weber received his doctorate in 1951 from the Catholic University of Washington, D.C. He was attending a seminar in Washington given by Gerard Herzberg of Canada one day while he was working on his thesis. Herzberg discussed stimulated radiation in his talk, and Weber realized that an amplification device could be made using stimulated waves. He published a paper outlining his ideas. In particular, he pointed out that a *population inversion* was important in designing such a device. A population inversion occurs when a high energy level contains more particles than a lower level. If heat or some other form of energy is applied to a system, many of the atoms become excited and are raised to upper energy levels. In theory, if you apply enough energy to the system, the levels will become equally populated. What is needed for a population inversion, however, is a level that is more heavily populated than a level below it (see fig. 15). For this to occur, "pumping" is required; in other words, energy of exactly the right frequency has to be injected into the system.

THE FIRST MASER

The first experimental operating maser was built by Charles Townes and a group of researchers at Columbia University. Interestingly, Townes's major interest when he entered Furham University in South Carolina was not physics; it was languages, and he received a B.A. in languages in 1936. But he took some physics courses, and a year later he received a B.Sc. in physics. In 1939 he obtained a doctorate from the California Institute of Technology. From there he went to Bell Labs where he worked on radar for the next few years. As a result of this work, he became interested in microwave spectroscopy, and in 1947 he left Bell for Columbia University where he set up a group to study microwave spectroscopy and the interaction between matter and microwaves.

Charles Townes

One of his major goals at this time was to produce a millimeter wave generator. He worked on it for several years, finally becoming frustrated with his lack of success. He was attending a conference in Washington, D.C., one day in spring 1951 when he began to think about the problems and why he had been unsuccessful.[5] Rising early the following morning, he went to a nearby park to contemplate the problem. Many attempts had been made to develop a millimeter wave generator using electronics, but he soon began to realize that it was best to go directly to molecules. He began making some rough calculations and was soon convinced that such a device could be made using the properties of molecules. He decided that the best way to proceed was to separate high-energy molecules from low-energy ones and then send them into a *resonant cavity* (a "box" with reflecting walls) where "stimulation" could be applied.

When he returned to Columbia, he and several colleagues began to look into the idea further. Among his colleagues was H. J. Zeiger, a postdoc fellow, and Jim Gordon, who was working on his doctorate. The molecule Townes finally narrowed in on was ammonia in the form of a gas. In order to understand the details of their device, we have to take a quick look at masers and lasers. Both devices come in two forms: as *amplifiers* and as *generators* or *oscillators*. Amplifiers receive an input signal and produce an output signal that has a greater energy than the input signal. Generators, or oscillators, accept energy in one form and convert it to another. In practice, an oscillator is usually an amplifier with a feedback mechanism so that some of the output energy is applied back to the input. This is where the cavity mentioned above comes in. As it has turned out, masers are usually amplifiers and lasers are usually generators.

The first thing that is needed for a maser (or laser) is a population inversion. This means that a way must be found to "pump" electrons up to a higher energy level so the atoms are in an excited state. This pumping can be accomplished in several different ways, and they will be discussed as we look in detail at the various devices.

The amplification principle can be best understood by looking at an energy level diagram.[6] Assume we have two energy levels, E_1 and E_2, as shown in figure 15. We pump electrons from level E_1 up to energy level E_2 until it is well populated. If E_2 becomes more heavily populated than E_1, we have a population inversion. With photons of the proper frequency, we can now stimulate the electrons on level E_2 to drop to level E_1. In the process, photons will be given off, and these photons will have exactly the same properties as those that stimulate the process. Of particular importance, we will get more photons out than we put in, so we will have amplification; furthermore the output photons will all have the same frequency and will be coherent (see fig. 16).

In most cases this simple amplification is not enough. We can, however, keep the wave in contact with the active material for a longer time and create a feedback process. This is usually done using a resonant cavity. In this case the wave is bounced back and forth between the walls; at each reflection the wave grows in strength until finally it is self-sustaining.

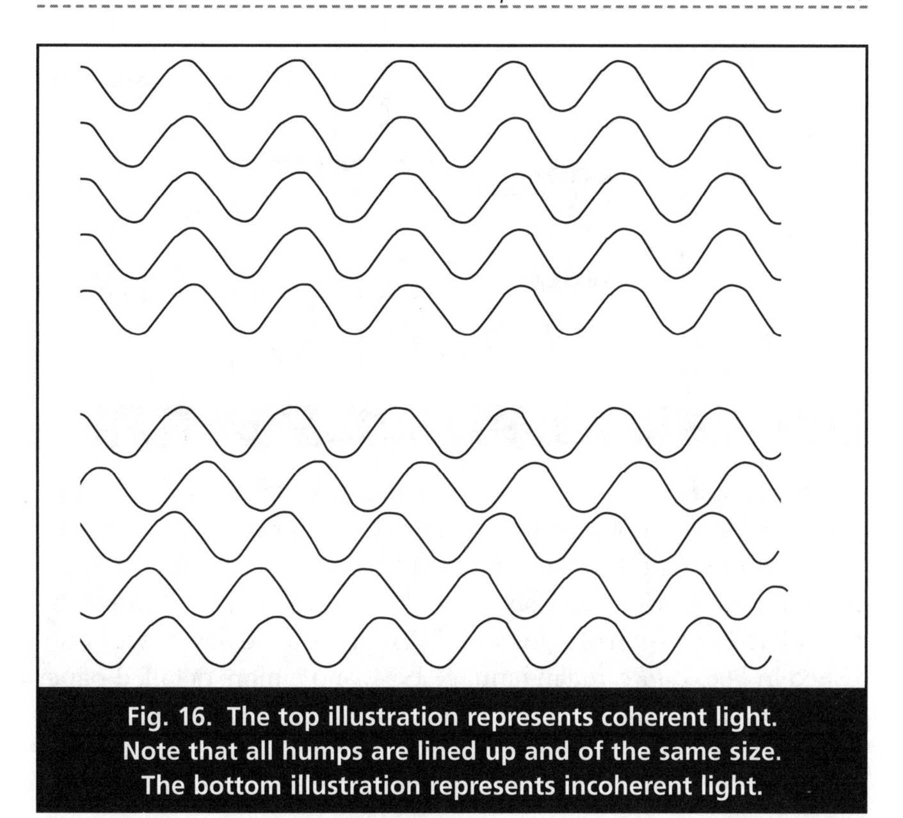

Fig. 16. The top illustration represents coherent light. Note that all humps are lined up and of the same size. The bottom illustration represents incoherent light.

Let's return now to Townes's maser.[7] He used ammonia gas as the active material. He started with a beam of ammonia that contained molecules in both an upper (excited) energy state and a lower state. These molecules were passed into an electric field where the low-energy ones were deflected away, and the remaining ones were channeled into a resonant cavity. All the molecules in the cavity were, therefore, initially in the excited state, giving rise to a population inversion. These molecules were then stimulated to fall to the ground state giving off coherent microwave radiation.

Townes and his students worked on the project for two years. Townes was worried at first that nothing would come of it and was advised at one point by two colleagues to give up the project. By that time he had spent $30,000 of government money on it. Finally, one day in 1953 Jim Gordon rushed into a seminar Townes was

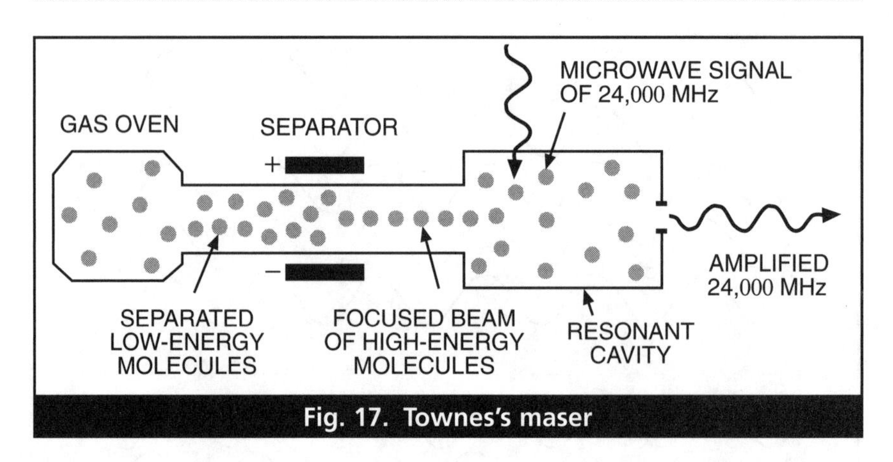

Fig. 17. Townes's maser

attending and yelled, "It works!" Townes was delighted and immediately rushed off to a nearby restaurant with his students to celebrate. A couple of days later they had coined the acronym MASER for the new device; it is short for "Microwave Amplification by Stimulated Emission of Radiation." A report on the device was published in *Physical Review* in January 1954, and a more detailed paper appeared a year later. The full realization of the potential of the maser, however, would not come for another twenty years. Townes received the Nobel Prize in 1964 for this and his work on the laser.

THE SOVIET MASER

While Townes and his students were working on the maser in the United States, two Soviet scientists were working on the same idea in the USSR.[8] Alexander Prokhorov and Nikolai Basov of the Lebedev Institute in Moscow published a theoretical paper on the maser about the same time as Townes. Their first working maser, however, did not come until a few months after Townes's model.

Prokhorov graduated from Leningrad University in 1939 and went to work at the Lebedev Institute. His career was delayed by World War II, but after the war he returned to the institute. Basov graduated from the Moscow Institute of Mechanics, and in 1948 he began work as a laboratory assistant at the Lebedev Institute. Soon he was collaborating with Prokhorov.

In their initial work, Prokhorov and Basov considered a two-level maser similar to Townes's. In November 1954 they published a paper in which they proposed a three-level maser using a gas as the active material. In this maser there were three energy levels, and the energy that was pumped into the system induced transitions from E_1 up to E_3. Since there were no electrons on level E_2, they had a population inversion. When a sufficient number of electrons accumulated on level three, it was stimulated, causing electrons to fall to level E_2, and the radiation that was given off was coherent. Pokhorov and Basov shared the 1964 Nobel Prize with Townes for their work.

A solid-state version of this three-level maser was proposed in 1956 by Nicolas Bloembergen. Born in the Netherlands, Bloembergen received his doctorate in 1943 from the University of Utrecht. He eventually came to the United States and worked at Harvard University. Although he proposed the device, he did not make a working model; it came in 1957 at Bell Labs when H. Scovil, G. Feher, and H. Seidel built the first three-level solid-state maser. Bloemberger received the Nobel Prize in 1981 for his work.

After 1957 numerous masers were built in laboratories throughout the United States. Ruby was shown to be a good material for masers in 1958, and it was employed in many of the masers that were built.

APPLICATIONS OF MASERS

Although it took many years to fully appreciate masers, they have now been used in many different areas. Because of their low noise (similar to "hissing" in the audio region), they are particularly useful in radio astronomy. Maser beams have been reflected from the surface of Venus, providing us with valuable information about mountain ranges and other surface features on the planet. The maser has also been used in radar systems where low noise is required. Atomic clocks use masers, and, of course, they have been used extensively in microwave spectroscopy.

In recent years, however, applications of the maser have been surpassed by those of the laser, a similar device that operates in the visible spectrum.

THE LASER

Once the maser had been invented, it wasn't long before there was speculation about a similar device in the optical region of the spectrum.

The region next to microwaves in the electromagnetic spectrum is the infrared, but it wasn't of as much interest as the optical, or visible light, region beyond it. Townes, naturally, began to think of this region immediately after he invented the maser. Others also took an interest. Prominent among them was R. H. Dicke of Princeton University. Dicke obtained his Ph.D. from the University of Rochester in 1941 and joined the faculty of Princeton University in 1946. He is best known for his alternative to Einstein's general theory of relativity and his suggestion that a cosmic background radiation might exist. Early in his career, however, he also made important contributions to laser physics. Dicke discussed the possibility of a laser that employs mirrors in its resonant cavity at a meeting of the American Physical Society in 1953. Three years later he discussed the same concept before a meeting of the International Commission on Optics. He published his suggestion in the

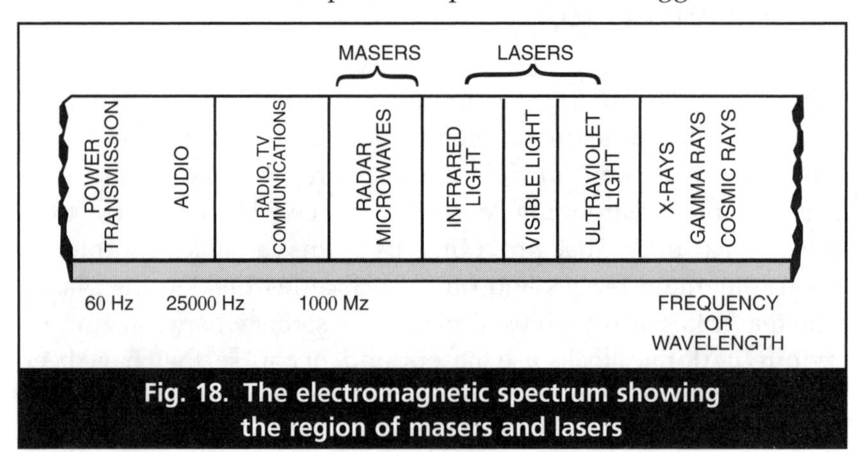

Fig. 18. The electromagnetic spectrum showing the region of masers and lasers

Journal of the Optical Society of America in 1953. Dicke called his device an *optical bomb* and predicted it would produce short, intense bursts of light.

Shortly after Townes began his work on the laser (at that time the laser was usually referred to as an optical maser), he teamed up with Arthur Schawlow. Schawlow obtained his Ph.D. from the University of Toronto, Canada, then went to work as a postdoc at Columbia University under Townes. In 1951 he left Columbia and went to Bell Labs. Townes was a consultant at Bell Labs, so the two men saw each other frequently.

In October 1957 they had dinner together. As they ate, they discussed possible models of an optical maser. By the time the meal was finished, they had decided to collaborate. Townes had already considered the possibility of an optical maser using thallium, and he sent some notes to Schawlow detailing his work. Schawlow showed that the idea would not work, so they quickly started searching for other materials. Within a short time, they had made considerable progress and decided to write a paper together. The major difficulty with the optical maser, as compared with the microwave maser, was the resonant cavity; the type used by the microwave maser would not work. Schawlow discovered, however, that two parallel mirrors placed some distance apart, with the active material between them, would work. A copy of the paper was sent to Bell Labs patent office so the device could be patented. Strangely, the office rejected the request, stating that optical waves were of no interest to Bell Labs and not part of their objective. Townes was annoyed and insisted it be patented, and in March 1960 a patent request was filed. The Townes-Schawlow paper was published in *Physical Review* in December 1958.[9]

Unknown to Townes, someone else was close on his heels. Another scientist had applied for a patent about the same time.[10] Gordon Gould had received his M.Sc. in 1943 from Yale and had gone to Columbia for his Ph.D. He was working in the Radiation Laboratory under Polykarp Kusch. Though his thesis project was not related to the optical laser, he had developed an interested in it and had made considerable progress. He had, in fact, arrived at the same idea for a cavity as Schawlow.

One day in October 1957 Gould received a call from Townes asking about a very powerful thallium lamp he had been working with in the lab. He knew Townes was also working on an optical maser, and the call got him excited. He was determined to beat Townes, and he knew that Townes was close. He immediately rushed out and had his laboratory manual notarized. Contained in it was the rough design of an optical maser. From there on in, however, he made several mistakes. First of all, he didn't publish his results. If he had, they might have appeared in the same volume of *Physical Review* as Townes and Schawlow's article. Second, he had spent so much time working on the device his thesis suffered, and Kusch did not approve it. He therefore left Columbia without receiving his doctorate.

Gould was offered a job at the research firm TRG Inc. Within a short time, the president of TRG learned of Gould's work and decided to take it on as part of TRG's research program. The research firm applied for a grant to work on the optical maser and eventually received almost a million dollars from the government. Gould filed for a patent in April 1959, which was about the same time Townes's patent was filed. Then TRG took Bell Labs to court over priority in the invention of the laser. As it turned out, Townes and Schawlow's patent was issued soon after they applied; Gould's, on the other hand, was not issued until eighteen years later. In fact, Townes and Schawlow's patent was just running out when Gould's came into effect. Gould then tried to collect royalties for the previous seventeen years, but most manufacturers fought him because they had already paid royalties to Townes and Schawlow. Eventually, however, Gould did collect some royalties.

THE SOVIET LASER

Gould was not the only one who was on Townes's and Schawlow's heels. V. A. Fabrikant and several students in the Soviet Union applied for a patent for a laser proposal in 1951, and it was issued in 1953. Fabrikant did not construct a laser, however, and his patent was based on a proposal. Basov and Prokhorov at the Lebedev Institute in Moscow had also started work on a laser. In 1958 they sent a proposal for a laser that used ammonia to the *Soviet Journal of Experimental and Theoretical Physics*. A year later Basov proposed a laser that used semiconducting materials; it was the first of many semiconductor lasers.

MAIMAN AND THE FIRST WORKING LASER

The first working laser was built in 1960 by Theodore H. Maiman of Hughes Research Laboratories in California using ruby as the active material.[11] Maiman earned his B.Sc. in engineering physics from Colorado University and his Ph.D. in 1955 from Stanford University. His area of specialization at Stanford was microwave spectroscopy.

Maiman's achievement of a ruby laser came as a surprise to most scientists working in the area. While Townes and Schawlow were concentrating on a gas laser, Maiman was working at Hughes on a ruby maser and was completely familiar with its properties. The heart of Maiman's laser was a cylindrical ruby rod a few centimeters long and about a half centimeter in diameter. One end of the rod was coated with a completely reflecting mirror and the other with a partially reflecting mirror. With these mirrors, the ruby itself was the resonant cavity.

Stimulated photons travel along the axis of the ruby, bouncing back and forth as they are reflected off the two mirrors at the ends of the rod. As they travel through the ruby, they stimulate more photons until eventually the beam is powerful enough to break through the partially reflecting mirror at the end of the ruby rod. A very intense flash lamp is wound around the ruby rod to pump electrons to the excited state. Because of the heat generated, the device had to be cooled by liquid nitrogen.

Maiman's paper was published in June 1960. It's interesting that his first paper was rejected by *Physical Review Letters*. They stated that they didn't think it had enough merit for rapid publica-

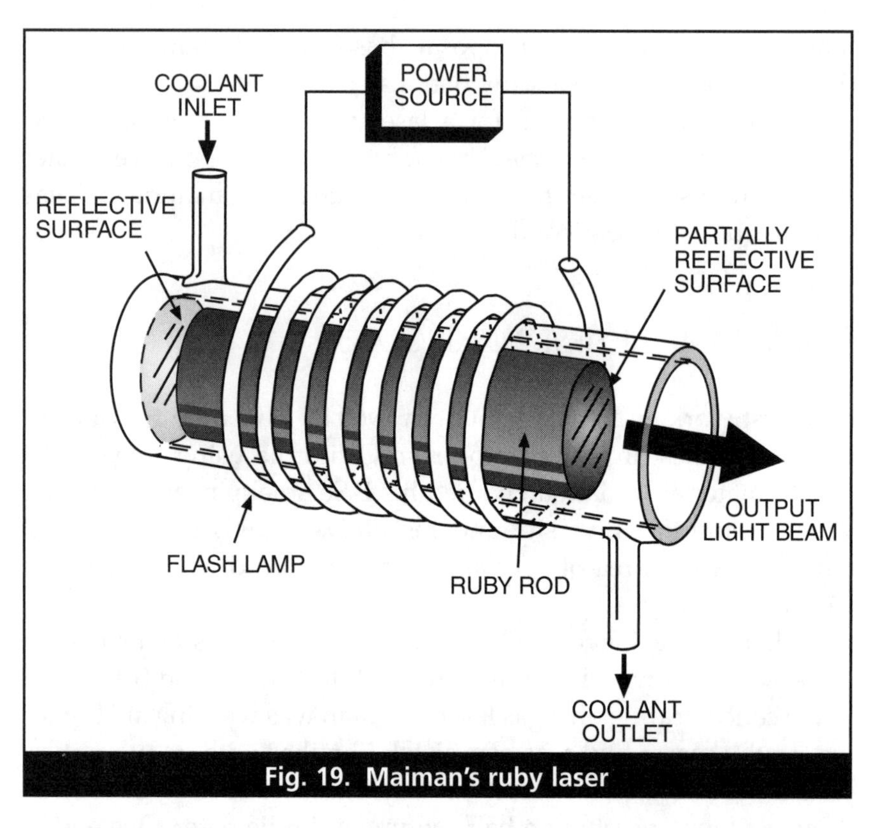

Fig. 19. Maiman's ruby laser

tion. Maiman then sent it to the British journal *Nature* where it was published immediately. Maiman's announcement created a race for devices based on other materials, and soon several were built.

THE GAS LASER

The first gas laser used a combination of the gases helium and neon, with about 90 percent neon and 10 percent helium. It was developed by Ali Javen of Bell Labs in 1959–60. The best way to understand this device is to look at the energy levels of the two gases. A radio frequency discharge is passed through the gas mixture to pump electrons from the ground state of helium to an excited level. In the process, some of the electrons become ionized; in other words, they break free from their atoms, and it is these

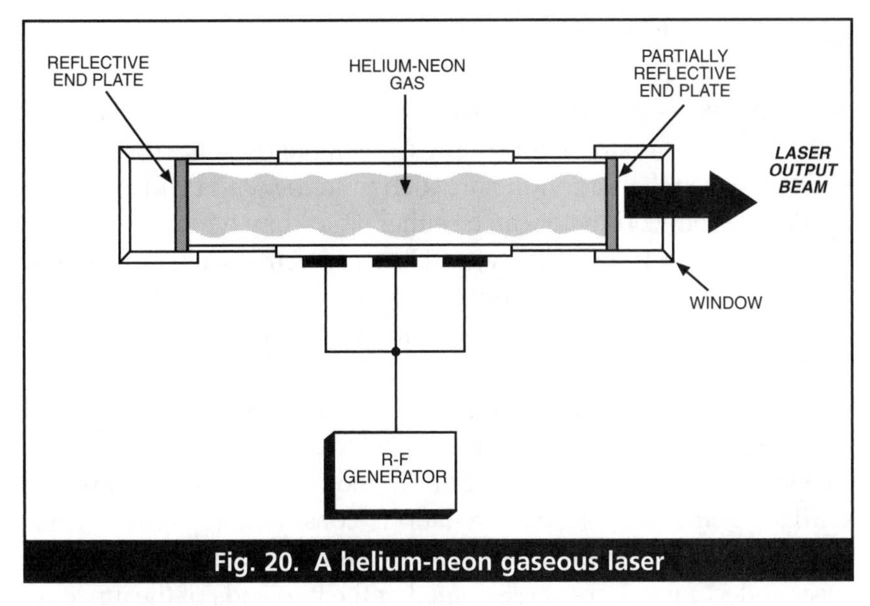

Fig. 20. A helium-neon gaseous laser

electrons that excite the helium atom via collision. But as we see in the energy level diagram, the excited level in helium is at the same position as E_3 in neon, so energy can be transferred from the helium to the neon. Energy levels E_3 (there are several of them very close together) soon become crowded, and a population inversion is created with respect to level E_2. Laser action takes place between levels E_3 and E_2 as electrons drop to E_2.

Although the helium-neon laser still dominates the field, numerous other gas lasers were developed after it. A mixture of nitrogen, carbon dioxide, and helium was used later. Mercury vapor has also been used, as has argon and several other liquids.

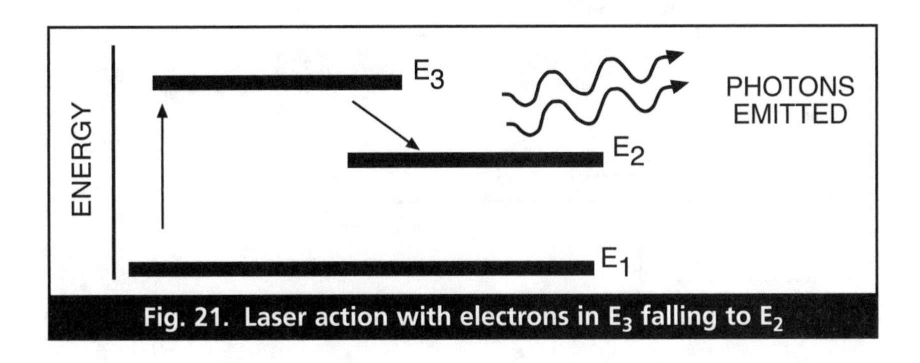

Fig. 21. Laser action with electrons in E_3 falling to E_2

SEMICONDUCTOR LASERS

We will discuss semiconductors in detail in the next chapter.[12] For now you can think of them simply as materials that have a conductivity between that of conductors such as copper and insulators such as glass. A semiconductor can be either *n* type or *p* type; *n* types have an excess of electrons and *p* types a lack of electrons, or equivalently, an excess of "holes." A *p-n* junction, which is constructed by joining *p*- and *n*-type semiconductors, acts like a diode in that it allows electricity to flow in only one direction. With the proper arrangement and enough current, large numbers of electrons and holes can be created near the junction; this is referred to as the *active region*. When an electron recombines with a hole, a photon is emitted, and with a sufficient amount of recombination, considerable light is produced—coherent laser light.[13] As in the case of the neon-helium laser, reflecting surfaces are situated at the two ends of the junction, as shown in figure 22, to enhance the beam.

The earliest proposal for a semiconductor laser was made by John von Neumann in September 1953. He wrote, "The possibility of making a light amplifier by use of stimulated emission in a semiconductor is considered. By various materials . . . it is possible to upset the equilibrium concentration of electrons. . . . The rate of radiation may be enhanced by incident radiation of the same frequency in such a way as to make an amplifier."[14]

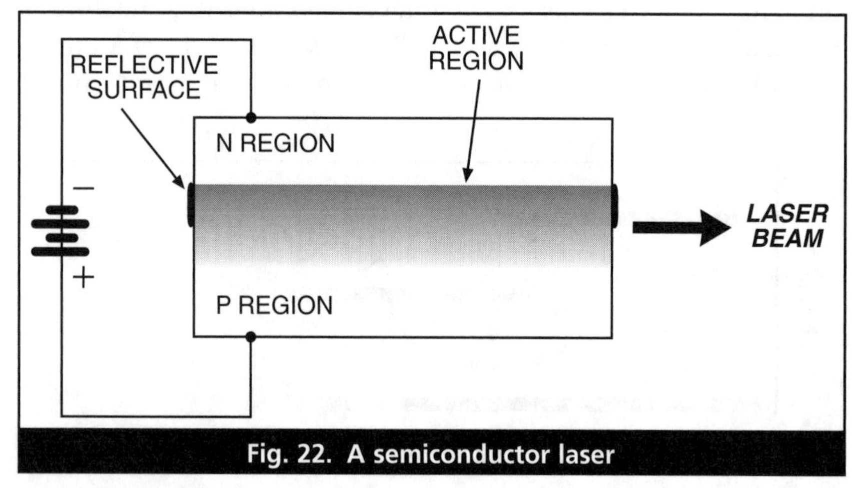

Fig. 22. A semiconductor laser

Many different types of semiconductor lasers have been devised in recent years. They have several advantages over other types of lasers. They are highly efficient compared to most lasers and can be operated at room temperature. They are also less expensive than other types.

MODERN APPLICATIONS OF LASERS

Without a thorough knowledge of energy levels gained through quantum mechanics, we would not have lasers or masers. The two devices, particularly lasers, are now used in an amazing array of applications. The following description of some of these uses is far from exhaustive.[15]

Beginning with ophthalmologists, we find that they use them to repair detached retinas, to destroy tumors in the eye, to repair broken blood vessels, and to place artificial lenses in the eye in cataract surgery. Surgeons use them to cut out tumors. It is known that some cancer cells readily absorb certain dyes. In some cases these dyes absorb laser light and can be heated and killed by the absorption.

Long flexible tubes with laser light at the end, called *endoscopes*, are now used routinely in medicine for exploratory purposes. The tube can be inserted into the mouth or nose to check for bleeding ulcers, burst blood vessels, or tumors. An endoscope can also be used to examine the lungs and large intestine.

A common problem in many older people, and some younger ones, is clogged arteries. Plaque buildup frequently cuts the blood supply to the heart, which can lead to a heart attack. Although the laser has not yet been used extensively in the removal of this plaque, it obviously has tremendous potential and will no doubt be used in the future.

Acupuncture, which is usually performed with needles, is also now being performed with lasers. The needles are replaced by short pulses from a laser. It is generally considered to be less painful and poses less risk of infection.

Dermatologists also now use lasers routinely. Lasers easily remove skin lesions, and they have also been used in the treatment of various kinds of skin cancer. The digestive tract is also easily reached by lasers. Lasers are also now being used to remove lesions, ulcers, and tumors in this region. Urologists now use the instrument for removing bladder tumors. Lasers may eventually be used against AIDS. They have been found to be effective in eradicating certain types of viruses in the blood. The procedure uses dyes and is similar to the one in which cancer cells are killed.

Although there's little doubt that the laser has become the "magic wand" of medicine in recent years, it also has important applications in the communication field. Laser beams are well suited for propagating messages over long distances because they do not spread out and dissipate as ordinary light does. Furthermore, because light has a much higher frequency than radio waves, more messages can be superimposed on the waves, and as a result a tremendous amount of information can be transmitted using a laser beam. Lasers have become particularly useful with the advent of fiber optics. Transparent, glasslike materials can be drawn into "wires" similar to the copper wires used in electrical communications. These wires can be bent just as copper wires can, and they can be insulated from one another in the same way. The laser light containing the signal is fed into one end of the fiber-optic cable, and it easily passes along it with little loss of signal.

Lasers are also now being used routinely for surveying. Measurements of distance are particularly important in surveying, and lasers provide accuracy that is impossible with other devices. Many other industries have also benefited from them. They are used extensively for cutting metals and welding. Not only can they cut metal, but they can also cut paper, plastic, and even cloth.

Most people have heard of lasers in relation to warfare, and a tremendous amount of research is now underway trying to devise lasers that can perform various war functions such as shooting down warheads. Laser weapons may one day be mounted in satellites, and it is possible that lasers will eventually be used to detect incoming missiles, and they may be used to illuminate targets so that guided missiles can be fired at them. It is also possible that soldiers will one day be equipped with laser guns and laser target locators, something that harks back to science fiction.

There are, indeed, numerous applications of lasers in warfare, but the applications for peaceful uses far outnumbers them, and none of them would be possible without the discovery of quantum mechanics.

Transistors and Superconductors

Of all the innovations derived from quantum mechanics, the one that has affected our lives the most is, without a doubt, the *transistor*. Its invention was truly a landmark in physics. Almost all modern electronic devices contain transistors in one form or another. They are now the tiny components of microcircuits in computers, radios, televisions, and hundreds of other devices.

To understand the role quantum mechanics played in their origin, we have to start with solid-state physics, a branch of physics that is in many ways an extension of quantum mechanics. Let's begin by considering a gas. The atoms of a gas are so far apart that the energy levels of a given atom are the same as those of a single atom of the same type. In other words, they are discrete. If we apply pressure to the gas, or lower the temperature, the atoms will begin to approach one another and eventually the gas will liquefy. The energy levels at the liquid stage are still distinct, but if we continue bringing the atoms closer together, the liquid will eventually solidify; at this point the energy levels overlap (fig.

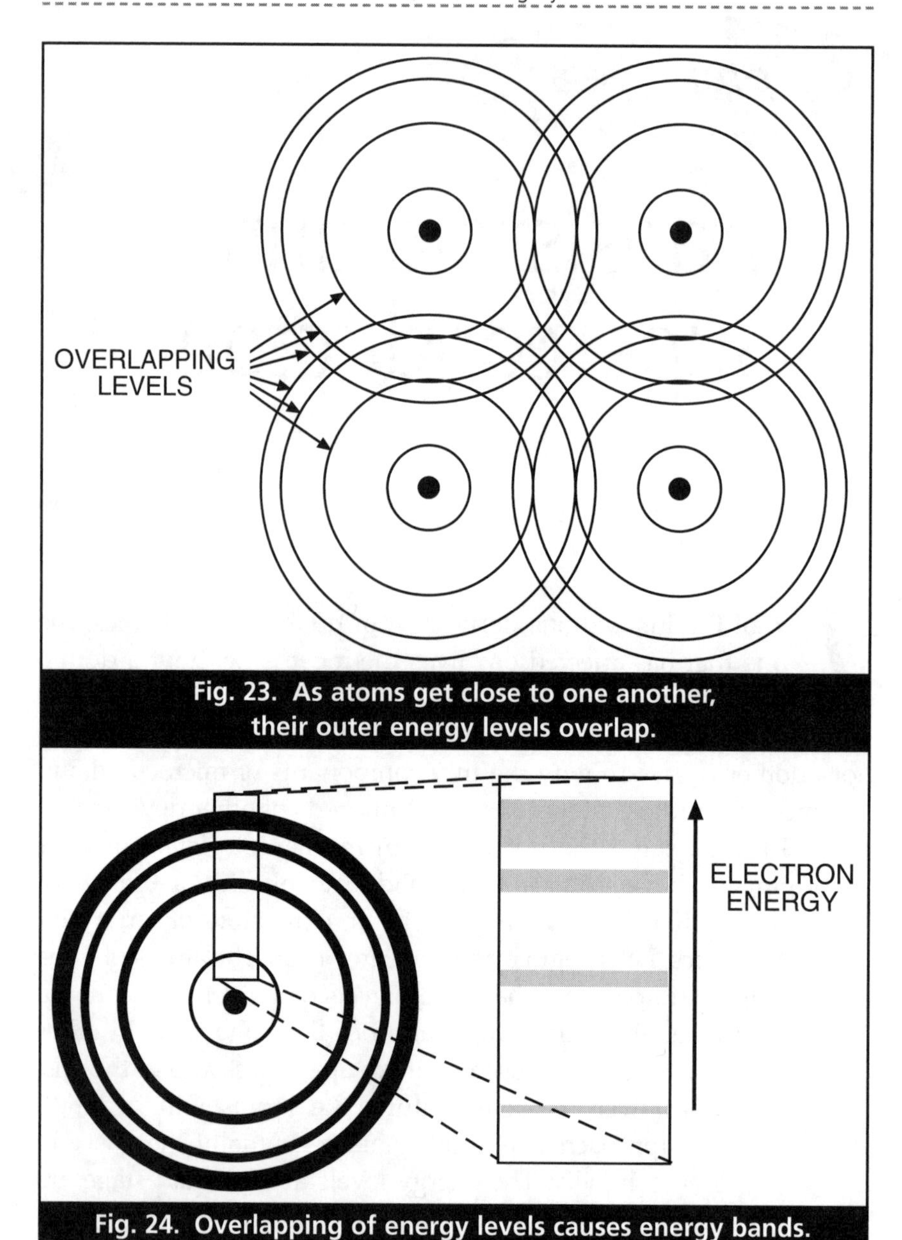

Fig. 23. As atoms get close to one another, their outer energy levels overlap.

Fig. 24. Overlapping of energy levels causes energy bands.

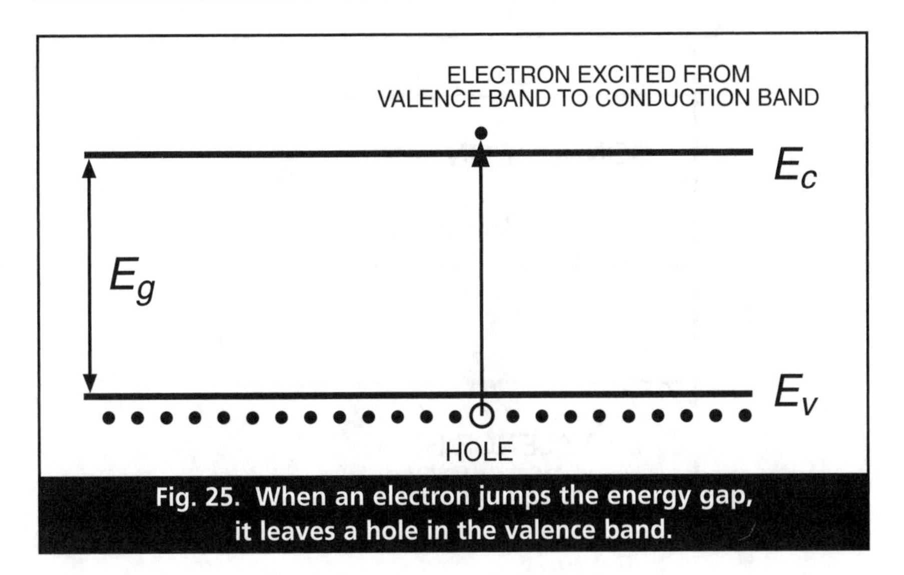

Fig. 25. When an electron jumps the energy gap, it leaves a hole in the valence band.

23) and create energy bands (fig. 24), or continuous regions of energy, separated by gaps.

Depending on the particular material, there will be several energy bands, and several gaps, and some of the bands will contain electrons. Furthermore, just as atoms have their lower energy levels (or ground state levels) filled at normal temperatures, with only a few electrons in the higher levels, so, too, will some of the bands be filled, and some empty. It is the difference between these bands and their structure in various materials that give us conductors, insulators, and semiconductors.

There are two energy bands of particular importance called the *valence band* and the *conduction band*.[1] In an insulator they are separated by a large gap (see fig. 25).

If an electron is to move through the lattice, it has to overcome this energy gap. In other words, it has to absorb an energy equivalent to the energy of the gap, and this usually doesn't happen in insulators.

Semiconductors, on the other hand, have relatively small gaps, so that moderate heat drives electrons from the valence band into the conduction band (see fig. 26).[2] The electrons in the conduction band belong to the lattice as a whole, rather than to individual atoms, and they move through the lattice, jumping from atom to atom. If a

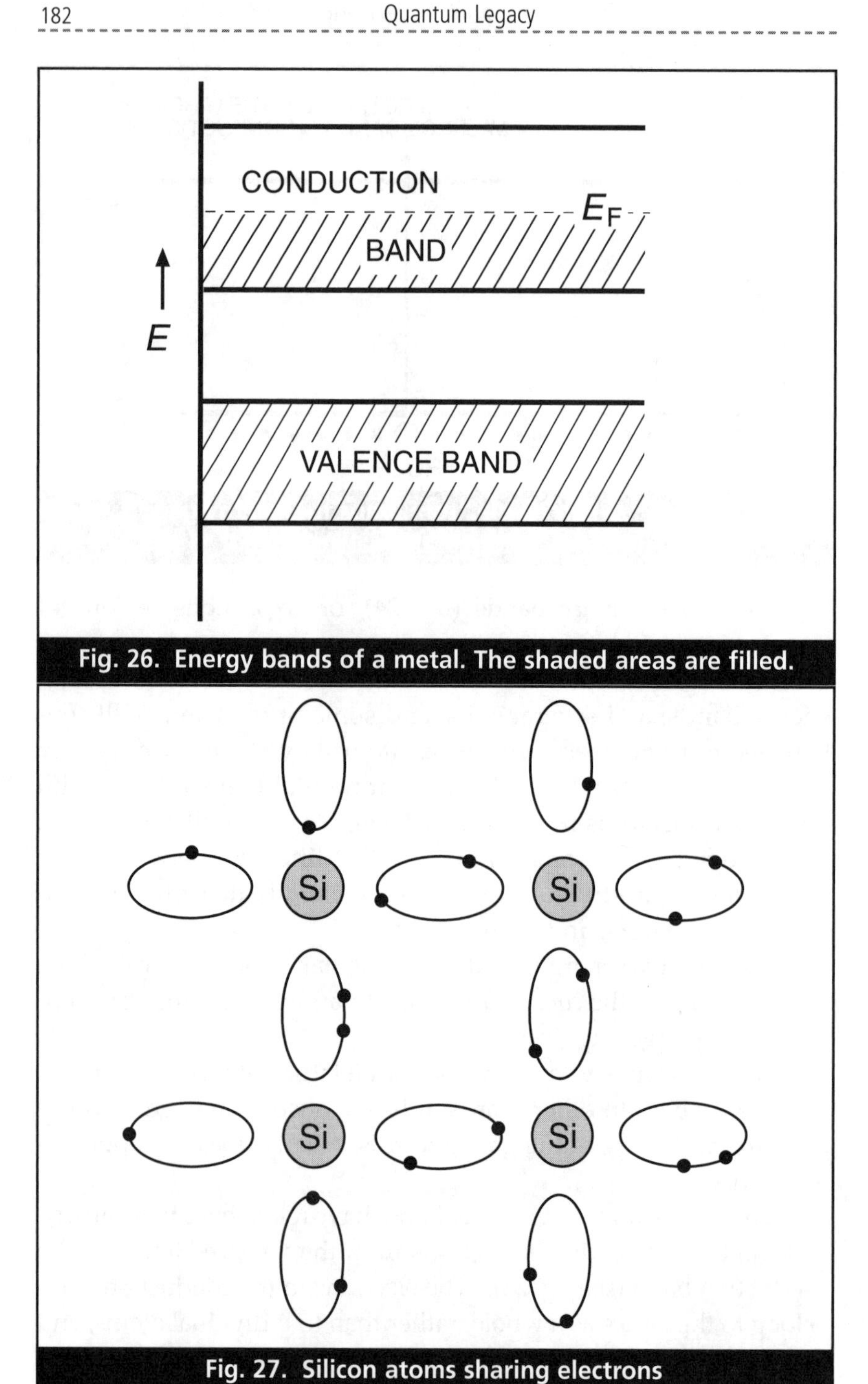

Fig. 26. Energy bands of a metal. The shaded areas are filled.

Fig. 27. Silicon atoms sharing electrons

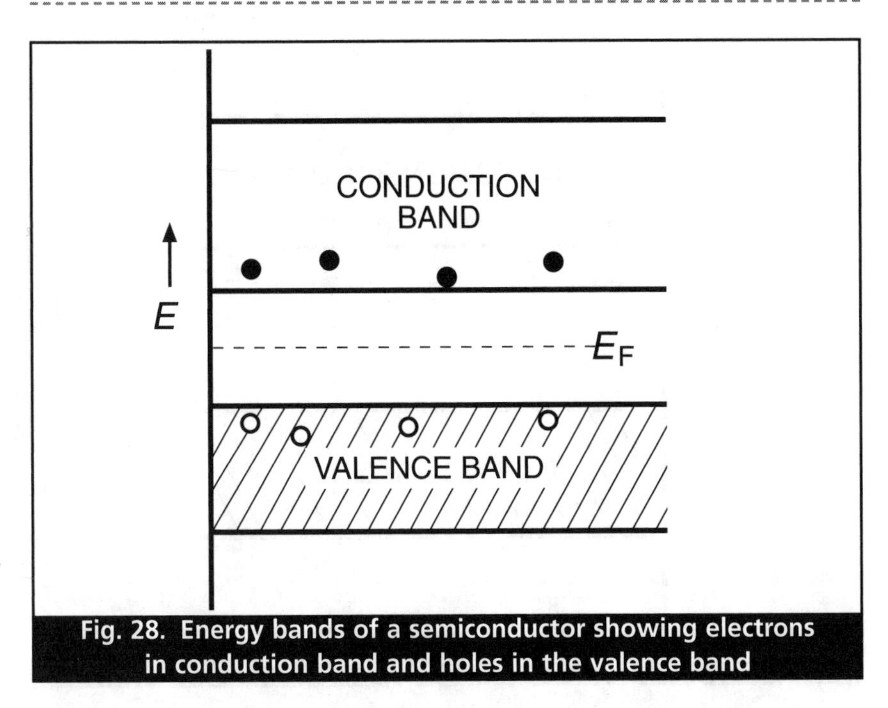

Fig. 28. Energy bands of a semiconductor showing electrons in conduction band and holes in the valence band

voltage difference (a battery) is applied to the semiconductor, the electrons will move in the direction of the positive terminal.

Finally, conductors such as copper and silver have energy bands that are overlapping. In this case there is a partially filled band and no gap to jump. If a voltage difference is applied, electrons will immediately begin to flow, and a current will be formed.

Since we will be dealing with transistors, most of our interest will be directed at the material from which transistors are made, namely, semiconductors. As we noted, when energy is supplied, electrons can jump from the valence band to the conduction band. This energy can be supplied as either heat or light. When an electron jumps into the conduction band, it leaves a "hole" behind in the valence band, and this hole can also conduct. In fact, you can have a "current" of holes, the only difference from an electron current is that this current moves in the opposite direction. In other words, holes are attracted to the negative terminal of the battery.

Let's look at this more directly—within the atom itself. In a lattice of silicon atoms (silicon is a semiconductor), the atoms are held together by sharing electrons. This is illustrated in figure 27. If one

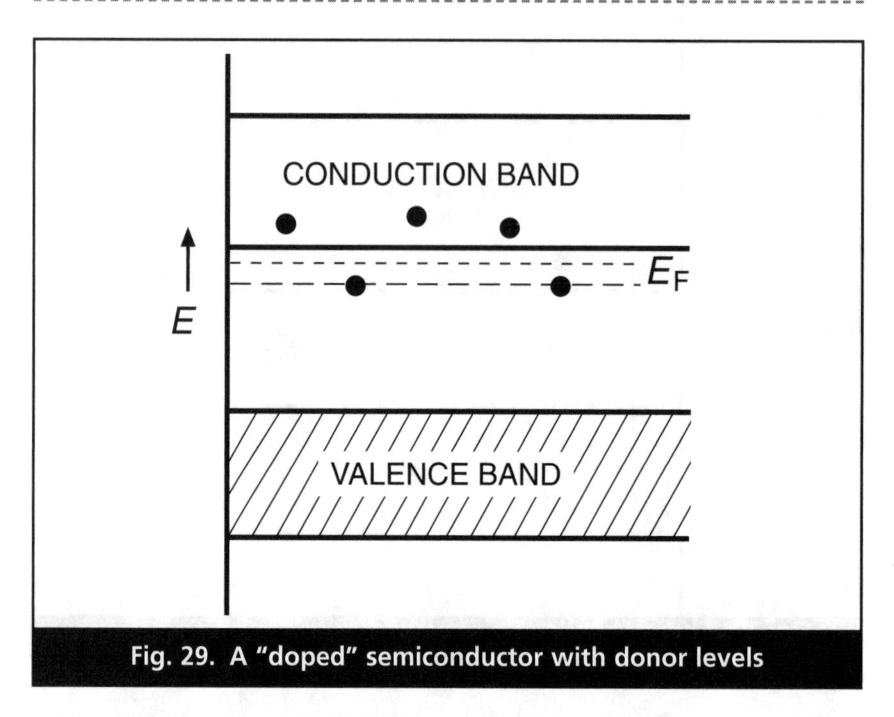

Fig. 29. A "doped" semiconductor with donor levels

of the electrons is missing, we have a hole, and just as electrons are free to wander from bond to bond, so, too, can holes.

Semiconductors such as silicon or germanium, which are made as pure as possible, are referred to as *intrinsic semiconductors*. In this case the number of electrons and the number of holes are equal (fig. 28). But we also have another type of semiconductor referred to as *extrinsic*. This type is made by "doping" the semiconductor with another atom, referred to as an *impurity atom*. In other words, we introduce a small amount of some other material into the semiconductor. This other material might, for example, be antimony. Antimony introduces new energy levels just below the conduction band, and since the electrons in these levels are only a short distance from the conduction band, they can easily jump into it (fig. 29). These levels are called *donor levels*, and the type of material is referred to as *n*-type. If, on the other hand, we dope silicon with indium, the energy levels in the gap will be just above the valence band, and they will act quite differently. We refer to them as *acceptor levels* because they accept electrons from the valence band, but

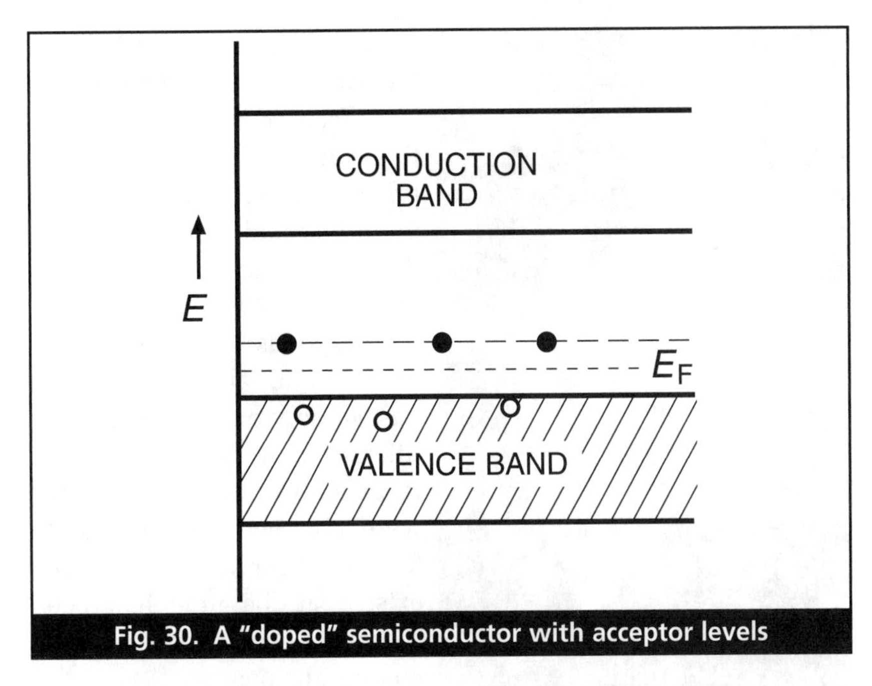

Fig. 30. A "doped" semiconductor with acceptor levels

when an electron leaves the valence band, it leaves a hole that can conduct. This type of semiconductor is referred to as *p*-type (fig. 30).

THE TRANSISTOR

With this, we're now ready to look at the discovery of the transistor. It took place at Bell Laboratories in Murray Hill, New Jersey, shortly after World War II, and is credited to three men: William Shockley, John Bardeen, and Walter Brattain. Shockley graduated from the California Institute of Technology in 1932, obtained his doctorate from MIT in 1936, and joined Bell Labs that same year. Bardeen graduated from the University of Wisconsin in 1928, obtained his doctorate from Princeton University in 1936, and joined Bell Labs in 1945. Brattain graduated from Whitman College in Washington in 1924, obtained his doctorate from the University of Minnesota in 1929, and joined Bell Labs in 1929.

The three men were the nucleus of a group that was formed in 1946 to study semiconductors—in particular, the semiconductors

John Bardeen

silicon and germanium.[3] Shockley was the leader of the group. They focused their attention initially on the phenomenon of *rectification,* which was important in radios and other electronic devices. It had been known for several years that certain crystals acted as rectifiers in that they would allow current to pass in only one direction. Rectifiers were of particular importance to the group because they were not well understood. In fact, the theory of rectifiers that existed at the time had been shown to make several incorrect predictions.

Prior to the formation of the group, Shockley had carried out a number of experiments in which he subjected semiconductors to electric fields. He had become convinced he could make an amplifying device using the effect, but when the experiments were carried out, the device didn't work, and he realized that there was still a lot to be learned about rectification and semiconductors.

After looking into some of the problems, Bardeen decided that the major difficulty was *surface states*—in other words, electrons that were on the surface of the semiconductor. He became convinced that it was difficult to obtain proper contact because of these surface states and decided that a more detailed study of them would have to be undertaken.

Of related interest was the *contact potential* of the surface. When a metal and a semiconductor came in contact, there was a readjustment of their energy levels, and the surfaces of the two materials become oppositely charged. This led to a potential difference between them. Brattain set up an apparatus to measure the contact potential of silicon. Since light can elevate electrons from the valence band to the conduction band, it also affected the contact potential. Brattain decided to measure the change in potential when either heat or light was applied to the sample. The first part of the experiment—using light as an energy source—was no problem, but when he looked into the changes that occurred at varying temperatures, he ran into a problem. Condensation kept forming on the silicon, and he had trouble getting rid of it. As he later wrote, "I toyed around in my mind. How can I do this experiment [in an easier way]—I was being lazy." He knew he could put the apparatus in a vacuum, but that was a lot of trouble. Another way to get rid of it was to dump the whole apparatus in water. It was a strange solution to the problem, but that's what he did. He measured the contact potential in water and then began checking other properties of the sample. To his surprise, he found that the sample acted like an amplifier. The water was somehow acting on the surface electrons, allowing amplification to take place. Brattain showed the effect to another scientist in the group, Robert Gibney. Together they examined the effect further, and soon there was no doubt: they had achieved a simple form of amplification.[4]

Walter Brattain

Amplification of a signal occurs when the signal is magnified or increased in intensity. There are three types of amplification: *current*, *voltage*, and *power*. In this experiment Brattain achieved current and power amplification, but not voltage amplification.

Brattain reported his results to the group at the next meeting. Bardeen was impressed and began thinking about the implications. The following day he went to Brattain's office and suggested that the sample need not be dipped in water. All that was needed, he was sure, was a drop of water and an electrode on the surface of the silicon. But there was a problem: the probe would have to be insulated from the water.

Brattain thought about it and quickly came up with an answer. He would cover the probe with wax; this would insulate it from the water. He would then push the wax-covered probe onto the silicon until it made contact. They tried it, and to their delight it worked. In fact, the drop of water worked just as well as when the entire apparatus was under water. Over the next few days, they tried several different variations on the setup and found they could get an even bigger increase in the current.

Then Bardeen suggested that they try *n*-type germanium instead of silicon. Brattain set up the experiment, and the two men watched as the current increased. Up and up it went, and when it stopped, they had obtained an amplification of 330. They were delighted. The current was in the opposite direction to what they expected, so it couldn't be an electron current. It was a current of holes; nevertheless, it was a current, and the amplification was beyond their greatest hopes.

But again there were problems. By the time everything was working well, the water drop had evaporated. Gibney suggested that they use glycol borate in its place. They tried the new material, which worked just as well, and evaporation was not a problem. But there was yet another problem: amplification was only obtained at low frequencies. This would not do. The device somehow had to be extended to include high frequencies. They decided to experiment further.

After looking into it, they concluded that the problem was the drop. But there was a catch-22. It was needed for amplification, but it worked only at low frequencies. They began experimenting with other probes, and during these experiments they noticed that there was an oxide layer under the drop of liquid, so Gibney proposed they prepare a sample with an oxide layer. In this case they might not need the drop. With the new sample prepared, they evaporated a spot of gold on the film so they could make contact with the surface. A tungsten electrode was placed next to it.

When everything was ready, Brattain began the experiment. As he probed he noticed there was no oxide layer beneath the probe. In washing off the glycol borate, they had accidentally removed the oxide layer. Brattain was annoyed, but he decided to continue with the experiment anyway. To his surprise, he got amplification. In fact, as they checked further they found that the amplification was at all frequencies—both high and low. The gold contact was acting like the drop in that it was producing holes in the germanium, and these holes were combining with the electrons on the surface in the same way that they had with the water drop. But this arrangement was much better.

Bardeen studied the results and soon realized a better arrangement of probes could be made. What they had to do was get two point contacts on the surface very close together. Bardeen calculated that they should be separated by only .002 inches. How could they get two probes so close? Again Brattain came up with a solution. He made up a small plastic triangle and covered the point with gold. He sent a current through the gold. Then, using a very thin razor blade, he made a cut at the apex of the triangle, separating the gold into two pieces. When the cut was complete, the current was broken. He then pressed it gently into the surface of the germanium. And it worked. They finally had what they wanted: an amplifier at all frequencies. It was the first point-contact amplifier (fig. 31). The two probes are now referred to as the *emitter* and *collector*. A third connection is made to the base of the germanium.

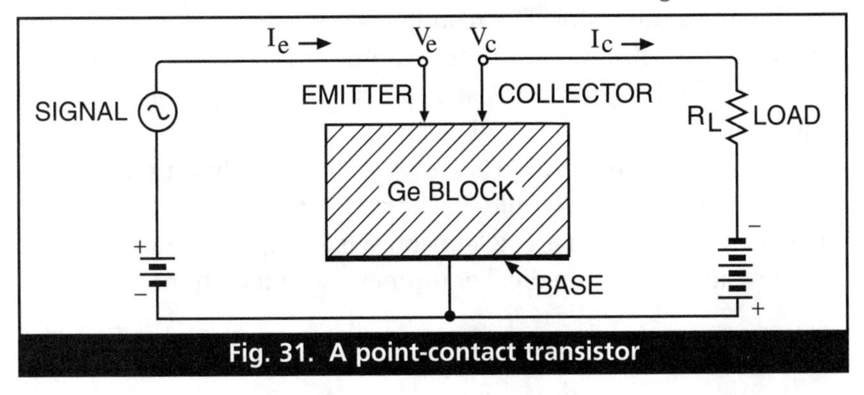

Fig. 31. A point-contact transistor

William Shockley

With this device, alternating current signals could be amplified in the same way as they were with the standard amplifying device of the time, namely, the triode tube. Transistors have a tremendous advantage over tubes, however, in that they can be made smaller, they operate at lower voltages, and they do not require a warm-up time as tubes do.

Bardeen suggested early on that the success of the point-contact transistor was due to a surface layer on the semiconductor. He believed the electrons on the surface restricted the layer. Experiments over the next few months indicated this may not be true. Shockley eventually became convinced it wasn't, and when John Shive placed emitter and collector points on opposite sides of a thin slice of germanium and still got transistor action, there was no doubt. This led Shockley to consider three layered structures, essentially two *p-n* junctions placed back to back. He noted that the diffusion of minority carriers (*p*-type in an *n*-type material and *n*-type in a *p*-type material) in the center region could be controlled, and he could make a transistor. The result was *n-p-n* and *p-n-p* transistors. The first successful transistor of this type was constructed in April 1950 (fig. 32). Junction transistors have now completely replaced point-contact transistors in all electronic devices.[5]

Transistors are used in so many different instruments and devices today that it would be hopeless to attempt to describe them all. Among them are: television sets, computers, radios, automobiles, appliances, medical equipment, and telephones.

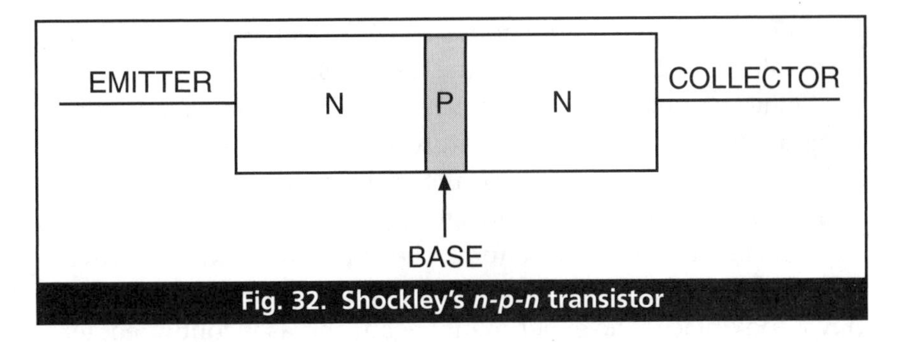

Fig. 32. Shockley's *n-p-n* transistor

SUPERCONDUCTORS

Although they share a similar name, superconductors are quite different from semiconductors. Like transistors, superconductors have also changed society in recent years, although not to the same extent. Nevertheless, they have become extremely important, and again our understanding of them has come to a large degree from quantum mechanics.

The discoverer of superconductivity, Kamerlingh Onnes, was born in Groningen, Holland, in 1853.[6] He attended the University of Groningen, receiving his doctorate in 1879. Upon graduation he went to the University of Leiden. From the beginning he was interested in low temperatures. Over several years he and his assistants built an apparatus that, according to his calculations, would allow them to get very close to the lowest possible temperature in the universe, namely, zero degrees absolute. This scale is also referred to as the Kelvin scale and degrees are in Kelvin.

Kamerlingh Onnes

Zero degrees Kelvin corresponds to –459 degrees on the Fahrenheit scale and –273 degrees on the Celsius scale.

Onnes completed his apparatus in 1908 and was ready for an assault on the lowest temperature ever attained by man. The experiment began on the morning of July 10, 1908. His, and his assistants', goal was to get down to approximately 4° K, which was the liquefaction point of helium. The experiment was given considerable publicity, and many visitors came to Onnes's lab on the day it took place. The experiment began, and as it continued, temperatures went lower and lower, but no liquid helium was seen. They pushed the apparatus, but it would go no lower. They knew they were very close to 4° K and should have observed it, but they saw nothing. Finally, at about 7:30 in the evening, they were ready to give up. As they went to shut down the equipment, one of the visiting professors suggested that they take another careful look. He cautioned that liquid helium would likely be very difficult to see. They moved the light around near the container that was being used to collect the helium, and, sure enough, about 50 cc had collected in the bottom of it.

Delighted with their success, they decided to see if they could solidify the liquid helium, but by 10 P.M. they had not been successful and decided to quit. They were successful, however, in obtaining the world's first beaker of liquid helium, and over the next few months, Onnes and his coworkers began looking into the properties of the strange new liquid. One thing that perked their interest was the resistivity of metals at low temperatures. Onnes was convinced that metals such as gold, silver, platinum, and mercury would have relatively low electrical resistances at such temperatures. He decided to begin with mercury. It could be obtained in a very pure state. He placed the mercury sample in liquid helium and recorded its resistance to electrical current as its temperature fell. As expected it decreased, but suddenly, to his surprise, at a temperature near 4.2° K, it dropped to zero. This meant the sample was presenting no resistance whatsoever to current. It was amazing because it meant that if a current was set up in a circuit made of mercury at this temperature, it would flow forever.

Onnes did, indeed, set up a circuit in which he used a battery to initiate current, and when he removed the battery, the current continued to flow for months. Later experimenters set up similar experiments in which the current flowed for several years. Onnes wasn't sure the resistance was exactly zero, but, as close as he could measure, there was no observed resistance. He referred to the phenomenon as *superconductivity*.

Onnes became very excited about superconductivity and predicted that many useful devices would one day be constructed as a result of it. In particular, he pointed out that powerful superconducting magnets could be made. But he soon discovered that there were difficulties. First of all, he noticed that there was a limit to the current the superconductor could carry before it reverted to ordinary conducting. This limit is now referred to as the *critical current*. Furthermore, he noted that the superconducting state was destroyed by a magnetic field, and the magnetic field (called the *critical field*) that could do this was not very large. With problems such as these, the prospect for superconducting magnets in the near future didn't look promising.

For several years superconductivity was a scientific curiosity, but no serious progress was made, and its potential seemed limited. One of the problems was that it was not understood. No one knew why a metal became superconducting; in other words, there was no theory of superconductivity. Finally in 1933 a breakthrough was made, this time by the German physicists Walther Meissner and R. Ochsenfeld. They discovered that a superconductor repelled magnetic fields; in other words, it could not be penetrated by a magnetic field. Although the breakthrough would play an important role in later developments, it did not lead directly to a theory.

SUPERMAGNETS—FINALLY

Superconducting magnets did eventually come, but it would take many years. One of the men responsible for their development was John Hulm. Hulm obtained his doctorate from Cambridge

University in 1949 with a thesis on *ferroelectrics* (the electric equivalent of permanent magnets). His thesis project consisted of growing ferroelectric crystals and measuring their properties. He was having trouble growing one particular ferroelectric, called barium titanate, when he heard of a technique that had just been developed. After receiving permission to use the technique, he grew some crystals, measured their properties, and published a paper on his results.

Bernd Matthias saw the paper and flew into a rage. Hulm had used his technique and had not asked for his permission. As it turned out, Hulm had asked only Matthias's supervisor and he, in turn, had not informed Matthias. Matthias took his anger out in a letter, but even that didn't help. He continued to seethe for months.

Soon after he completed his work, Hulm had an opportunity to come to the University of Chicago, and he took it. Shortly after arriving in Chicago, he was invited to a dinner party at the home of his supervisor, Andrew Lawson. During the evening Lawson introduced him to several of the people at the party. Among them was Matthias. Matthias eyed him as the two men were introduced.

"Are you from Cambridge?" he asked.

Hulm replied that he was.

"You're the guy that stole my work on barium titanate," he said in a loud, angry voice.[7]

Hulm didn't know how to reply and tried to calm him down by telling him that he did get permission. But it didn't help. Matthias raged on, eventually challenging him to a duel. Hulm managed to escape when dinner was served, and to his relief he wasn't seated near Matthias.

After the dinner Matthias approached him again, but now he was calmer. He asked what he was working on now and Hulm replied that he was still working on ferroelectrics. Soon they were deep in conversation about their work and all animosity was forgotten. By the end of the evening, they had actually agreed to collaborate, and over the next six months, they worked together measuring the properties of many different ferroelectrics near absolute zero. After they had accumulated considerable data, Hulm gave a seminar to the faculty and students of the University of Chicago. In the audience was Enrico Fermi, the Italian American physicist who obtained the first sustained fission reaction. Fermi complimented them on their work and then asked why they were spending so much time on the superconductivity of ferroelectrics when almost nothing was known about superconductivity itself.

Matthias and Hulm thought about Fermi's remark and decided he was right. They would redirect their research to the underlying mechanism behind superconductivity. The question of why certain materials became superconducting while others did not was obviously important, and it was a challenge to them. They began looking at all the materials that exhibited superconductivity for a clue. During the process they discovered large numbers of new superconductors and became experts at predicting which materials would be superconducting.

Despite the progress Matthias was making and the papers he had published, the University of Chicago refused to promote him or give him tenure. He eventually became dissatisfied and left for Bell Labs. The two men, however, kept in touch.

Not long after arriving at Bell Labs, Matthias was talking to a coworker, Rudi Kompfner, when the conversation turned to superconducting magnets. Some progress had been made in finding superconducting materials that could survive relatively strong magnetic fields, but it was still a problem. Matthias suggested that Kompfner try an alloy of molybdenum and rhenium. Kompfner had the alloy made up, and to his surprise it generated a relatively large field of 16,000 gauss. (A small bar magnet typically has a strength of 100 to 200 gauss.) Matthias then suggested several other alloys, and within a few months they had fields up to 88,000 gauss. The era of superconducting magnets had begun, and before long they were a common item in labs throughout America. Not only could they generate huge fields, but they could be made much smaller and more compact than conventional magnets.

THE BSC THEORY

A considerable amount was now known about superconductivity. Several hundred materials were known to exhibit the phenomenon, but there still wasn't an adequate theory to explain it. Many of the best-known theorists had tried their hand at formulating a theory, but no one had been successful. Then in 1950 a discovery was made that would play an important role in the formulation of a theory. Emanual Maxwell of the National Bureau of Standards and Bernard Serin of Rutgers University, independently, discovered what is now known as the *isotope effect*. (Isotopes of a given element have the same number of protons but a slightly different number of neutrons.) Maxwell and Serin showed that when the transition, or critical, temperature of two isotopes of the same element were measured, the heaviest isotope always had a slightly lower transition temperature. When John Bardeen (the coinventor of the transistor), who was now at the University of Illinois, heard about the effect, he was sure he could use it to formulate a theory of superconductivity. Over the next few years, he worked on a theory but made little progress. Then he began a collaboration with a young physicist, Leon Cooper.[8] Cooper had received his doctorate from Columbia University in 1954 and was at the University of Illinois as a postdoc. Soon after Cooper began working on the problem, he came to an important conclusion: the electrons in a superconductor somehow worked in pairs. In other words, they cooperated with one another. They are now referred to as *Cooper pairs*.

Even with this insight, however, the problem was far from solved. Quantum mechanics was needed to set up a theory, but serious difficulties remained. The phenomenon was known to be collective in that it resulted from the cooperative action of many atoms, and this kind of problem was difficult to deal with using quantum mechanics. Bardeen decided to assign certain aspects of it to Robert Schrieffer as a thesis project. Schrieffer shared an office with Cooper.

Schrieffer worked on the problem for several months and made no progress. He soon grew frustrated and eventually went to Bardeen and asked to be assigned to a different problem. Bardeen urged him to give it another month; if nothing developed, Bardeen agreed to give him a different project. Within a short time, Bardeen left for Stockholm to receive the 1956 Nobel Prize for his work on the transistor, and while he was gone, Schrieffer and Cooper continued to struggle with the problem. Suddenly, to their delight, several pieces of the puzzle came together. They reported their breakthrough to Bardeen as soon as he got back, and together the three of them worked to bring all of the pieces together. Within a short time, they had a successful theory. Now referred to as the *BSC theory*, after the initials of the three men, they presented it at an American Physical Society meeting in 1957 and published it in *Physical Review* that same year. They received the Nobel Prize for the theory in 1971, and Bardeen became the only physicist to ever win two Nobel Prizes.

HIGH-TEMPERATURE SUPERCONDUCTIVITY

Superconductivity normally takes place at temperatures very close to absolute zero. Different metals, alloys, and so on, however, have different transition temperatures. For several years the highest known transition temperature was 23° K. This was not surprising in that the BSC theory predicted that the highest transition temperature would likely be in the range 30 to 40° K. To most, it therefore seemed fruitless to search for materials with higher transition temperatures. But if liquid nitrogen temperature (77° K) could be reached, superconductivity would be much more accessible, and devices and innovations based on it would be more likely to be developed.

One person who was interested in the possibility of high-temperature superconductors was K. Alex Müller of Switzerland.[9] He received his doctorate from ETH in Zurich in 1958, worked for a few years at Battelle Institute in Geneva, then went to IBM in Zurich. Although he had published a few papers, he was little known outside of Switzerland. In 1980 he came to IBM in the United States for a year and a half visit. It was here that his interest

Alex Müller

in high-temperature superconductors began. By the time he returned to Switzerland, he had decided to search for high-temperature superconductors, and he was sure the "oxides" were his best bet—in particular, ceramics (a mixture of materials with nonmetallic properties, such as pottery—usually "fired" at high temperatures).

Earlier he had met Johann Bednorz, who was also a graduate of ETH. Bednorz had become an expert in growing exotic crystals and enjoyed experimenting with new combinations of various materials. Müller knew of his talents and decided to ask him if he was interested in collaborating on the high-temperature project. Bednorz was now at the IBM Lab in Ruschlikon. Müller visited him one day, prepared to give a strong argument for the importance of the work, but to his surprise Bednorz agreed to a collaboration almost immediately. He agreed to grow the crystals, and Müller would test their transition temperature. They also agreed to keep the project secret and only work on it part time.

Müller decided that oxides of either nickel or copper were the best bet, and over the next two years, they tried every conceivable combination of nickel they could think of. Finally they decided to switch to oxides of copper, but they still didn't get what they wanted. Two and a half years had passed when they decided to review their options before proceeding. It was 1985, and they were becoming discouraged. After so much work, they had found nothing that had a transition temperature above 23° K. Then one day Bednorz came across an article by a French chemist in which the properties of a ceramic consisting of barium, lanthanum, copper, and oxygen were discussed. The strange mixture immediately caught his attention, and he decided to grow a crystal of the material to see what its transition temperature might be.

Johann Bednorz

He made up a sample just before the Christmas holidays, and there was no time to test it before the holidays. It therefore sat in the lab for over a week, but when they finally got back, they immediately tested it. To their surprise the transition temperature was 35° K—twelve degrees higher than the old record. Müller could hardly believe it. Bednorz wanted to publish the result immediately, but Müller was cautious. He wanted to check everything thoroughly first. They repeated the experiment several times, and each time they got the same result. But there was one last thing they had to do to be sure it was a superconductor. They had to check for the *Meissner effect* (where superconductors expel magnetic fields from their interior), and they didn't have the apparatus for doing it.

It would take several weeks to get the equipment they needed to test for the Meissner effect, and it was too risky to delay publication. They might be scooped. They decided to publish in an obscure journal—a German journal called the *Journal of Physics*—and they would give the article an obscure title. Few people were likely to see it while they waited for their apparatus, yet it would establish their priority if there were problems. The apparatus finally arrived, and they checked for the Meissner effect. The test was positive! There was now no doubt: it was a high-temperature superconductor. They immediately wrote up another paper and published it, and in 1987, only one year later, they were awarded the Nobel Prize for their work.

THE RACE WAS ON

As it turned out, Müller and Bednorz were not the only ones looking for high-temperature superconductors. One of Matthias's former students, Paul Chu, who was now at the University of Houston, was also after the same goal. Like Müller, he was interested in oxides of various metals. He had determined that pressure increased the transition temperature and was "squeezing" his samples to see how high he could take them. He was disappointed when he saw Müller and Bednoz's article because he had come close to the ceramic they had discovered but had missed it. He was determined, however, to find something with an even higher transition temperature.[10]

Robert Cava and Bertram Batlogg at Bell Labs soon got into the race. Maw-Kuen Wu at the University of Alabama also joined it. Wu later joined forces with Chu, and the two groups began concentrating on compounds made up of yttrium, barium, copper, and oxygen. Wu's group hit the magic combination first. The transition temperature of the new mixture was 89° K. Wu immediately took the material to Chu's lab to test it with Chu's more sophisticated equipment, and soon there was no doubt.

On February 15, 1987, Chu announced that he, in collaboration with Wu and his group, had found a superconductor with a transition temperature of approximately 90° K. The news stunned the scientific world. Chu and Wu sent a paper to *Physical Review Letters* for publication. At this stage, however, they wanted to keep the formula secret, so they waited until the last moment to include it in the article. The annual meeting of the American Physical Society was coming up the following month, and there was a great deal of anticipation. On March 18, the day of the session, crowds began to assemble in the New York Hilton where the meetings were held. The crowd grew to almost four thousand, and there was only room for two thousand in the hall where the presentation was to be made. Closed-circuit television monitors were quickly set up. Later, the meeting would be referred to as the "Woodstock of Physics," and indeed there was much of the same frenzy and excitement.

Less than a year later, in January 1988, Zhengzhi Sheng and Allen Hermann of the University of Arkansas took the record even higher. They found a superconductor with a transition temperature of 125° K. Then in 1995 a team in the Midwest obtained one with 138° K. It is the current record.

One of the things that was particularly strange about the new discoveries was that the BSC theory predicted that the highest transition temperature should only be about 40° K. It was obvious that there was a problem with the theory. Several people have attempted to modify it; among the most successful has been Philip Anderson of Princeton University. In 1987 he came up with a modification of the theory that explained Cooper pairs, but Anderson's theory made no new predictions and was soon shown to be lacking. Others tried their hand, but we still do not have a completely satisfactory theory of the phenomenon. Nevertheless, a new and more satisfactory theory will eventually be found, and it will no doubt be based on quantum mechanics.

THE SPIN-OFFS OF SUPERCONDUCTIVITY

Superconductors don't quite fall into the same category as transistors when it comes to benefits to society; nevertheless, they are helping to improve it. Furthermore, quantum mechanics is critical to the understanding of the phenomenon although it hasn't played the same role it has in other areas. The reason is that superconductivity is particularly complicated, and even though the BSC theory is an excellent theory, there are things it doesn't predict. Still, we have had many important innovations from superconductors. The MRI scanners in hospitals that allow doctors to search for tumors depend on superconductivity. And most of the large accelerators around the world, which are used for extending our knowledge of the universe and its makeup, use superconducting magnets. The Tevatron at Fermilab, for example, is a 3.5-mile ring of superconducting magnets.[11] High-speed trains are now levitated by superconducting magnets. Japan leads the world in this aspect, but considerable progress is being made in other countries.

Another promising area is superconducting magnetic energy storage (SMES). Such a device would allow energy plants to store energy. At the present time, considerable energy goes to waste because it cannot be used when it is generated. Although several schemes have been devised for energy storage, they are not very efficient. Superconducting magnetic storage is expected to be much more efficient.

In 1962 an English physicist, Brian Josephson, showed that when two pieces of superconductor are separated by a layer of insulator, electrons can tunnel through the barrier under the right conditions. It is now referred to as the *Josephson effect*, and several devices have been built based on it. One of the most useful is a very sensitive detector of magnetic fields called a SQUID. It has both civilian and military applications. It can be used to detect submarines and mines, and it is also being used to map Earth's magnetic field. It may eventually be helpful in locating mineral deposits.

Superconductors may also play a role in computer technology one day. Research is now looking into the possibilities. Computer logic chips that use superconductors have been built and work reasonably well. Superconducting memory units have also been constructed, but they have not been as successful. Superconducting motors and generators are another possibility in the future. Needless to say, it's tough to predict the future, but the potential for applications seems boundless.

The Nuclear Age

No science depends on quantum mechanics more than *nuclear physics*, the branch of physics that deals with the nucleus of the atom. Almost immediately after quantum mechanics was discovered, scientists began applying it to the nucleus. Many of the initial attempts were unsuccessful because little was known about the nucleus. The neutron had not been discovered, and there were strong indications that there were electrons in the nucleus. Beta rays, for example, which appeared to come from the nucleus, were electrons.

Niels Bohr and his colleagues at Copenhagen tried unsuccessfully to formulate a theory of the nucleus in the early 1930s, and Bohr eventually became dejected over his lack of success. His pessimism spilled over to Heisenberg. But within a short time, a number of experiments led to a better understanding of the nucleus. In 1930 W. Bothe and H. Becher of Germany projected a beam of alpha particles at a thin sheet of beryllium and found that highly penetrating radiation appeared to be emitted from the beryllium.

They assumed it was gamma rays, but were confused by some of the results of the experiment.

The daughter of Marie Curie, Irène Joliot-Curie, and her husband, Frédéric Joliot-Curie, repeated the experiment in 1932, but they placed a sheet of paraffin beyond the beryllium. They found that highly energetic protons were ejected from the paraffin by the mysterious radiation that came out of the beryllium. They interpreted it as the emission of protons as a result of highly penetrating gamma rays. But this seemed unlikely to James Chadwick of Cambridge University in England. He was sure that gamma rays would have little effect on the proton, which was 1,836 times as heavy as the electron.

Chadwick had worked for Rutherford since 1919, bombarding various elements with alpha particles. For several years both he and Rutherford had wondered about the possibility of a neutral particle in the nucleus, but they had been unable to show that one exists. Chadwick was sure the Joliot-Curie interpretation was wrong and repeated the experiment. He then substituted several other elements for beryllium and compared the results with and without a paraffin sheet beyond the element, finally coming to the conclusion that it was not gamma rays that were knocking the proton out of the paraffin, but rather a neutral particle with a mass approximately equal to that of the proton. He called the new particle a *neutron*. His interpretation proved to be correct, and in 1935 he received the Nobel Prize for the discovery.[1]

The nucleus appeared to contain protons and neutrons. But did it also contain electrons? Heisenberg soon came to the conclusion that it did, but it played a strange role. In a paper published in July 1932, he applied quantum mechanics to the problem and obtained a new theory of the nucleus. He postulated that the neutron was made up of a proton and an electron, with the positive charge of the proton canceling the negative charge of the electron.

One of the major problems in earlier attempts at a model of the nucleus was the nature of the force that held the *nucleons* (particles of the nucleus) together. Heisenberg used the hydrogen molecule as his model. In this molecule the two hydrogen nuclei (protons) are held together by sharing an electron. Heisenberg visualized the nucleus as made up of protons and neutrons, held together by passing electrons back and forth.[2] When a neutron passed an electron to a proton, it became a proton, and the proton that received the electron became a neutron. The force was called an "exchange" force since it resulted from the exchange of electrons. His model was very persuasive in that it explained many of the major problems associated with the nucleus such as deuteron binding (a neutron bound to a proton), certain aspects of radioactive decay, and the stability of the helium nucleus.

There was, however, a difficulty. The "glue" that held the nucleons together had to be very short-ranged. It could only act over a distance about the size of the nucleus, and the nucleus was 100,000 times smaller than the atom. It was, in essence, like a tiny grain of dust at the center of the atom, yet it contained most of the mass. Three years later, Hideki Yukawa of Japan proposed that the electrons be replaced by what he called *mesons*, and a better understanding of the nuclear force followed. Heisenberg's theory, however, was an important step, and all of field theory is now based on the idea of an exchange force due to the exchange of particles.

The Joliot-Curies were disappointed that they missed the discovery of the neutron, but in 1934 their time came. They were bombarding light elements with alpha particles when they noticed that after the bombardment ceased radiation continued to be emitted. Looking into it further, they discovered they had produced an isotope of phosphorus, and of particular interest, the isotope was radioactive. It was emitting radiation and breaking down to a lighter element in the process. Until then radioactivity had only been observed in very heavy elements. For their discovery, a process we now refer to as *artificial radioactivity*, they received the Nobel Prize in 1935.[3]

Another of the major problems of the time was spontaneous decay of the nucleus, referred to as *alpha decay*. Heavy nuclei such as uranium emitted alpha particles, but there was something strange about the process. If you projected an alpha particle at the nucleus, it was deflected. There was a "barrier" around the nucleus, and it was particularly strong—so strong that scientists wondered how alpha particles could penetrate it from the interior. In other words, since it was well known that the nucleus emitted alpha particles, how did they manage to get out? The answer came from George Gamow, a large, fun-loving, practical joker, who never seemed to run out of energy. Born in Russia in 1904, Gamow's early education was frequently interrupted by war. Indeed, he was almost killed when a shell exploded just outside the school building he was in. It shattered the window he was standing near and knocked him off his feet. Despite the turmoil around him, he developed an early interest in physics and astronomy. His enthusiasm for astronomy was kindled when his father gave him a telescope for his fourteenth birthday. Fortunately, by the time he was ready to go to university, things had calmed down, and over the next few years he attended both Novorossia University and the University of Leningrad.[4]

Upon graduation from the University of Leningrad in 1928, he spent time at the University of Göttingen, at Neils Bohr's institute in Copenhagen, and at Cambridge University in England. It was an inspiring time in physics with everyone caught up in the excitement that surrounded the discovery of quantum mechanics, and Gamow wanted to share in it. It was evident to him that everyone was working in the same general area, so he decided to look for something new to work on. He finally narrowed in on the nucleus of the atom. It was an area about which very little was known. One thing that was not understood was *spontaneous decay*—the emission of alpha particles from the nucleus. Applying quantum mechanics, he showed that the alpha particles could

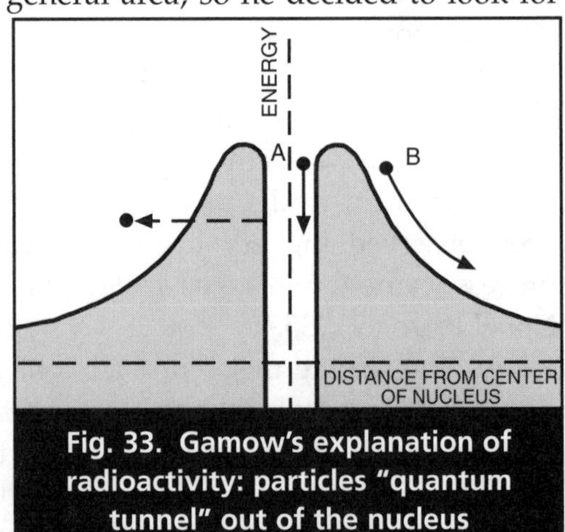

Fig. 33. Gamow's explanation of radioactivity: particles "quantum tunnel" out of the nucleus

"quantum tunnel" through the high barrier around the nucleus. His explanation centered on the uncertainty principle; he showed that according to it there was a small probability that the particles could "sneak" through the barrier, as long as they did it fast enough. Another way of looking at it is that the particle "borrows" the needed energy for a very short period of time. If it pays it back within an extremely short period of time specified by the uncertainty principle, no physical law is violated (fig. 33). It turned out to be one of Gamow's major contributions to physics and an important breakthrough in nuclear physics. Later, he also proposed a "liquid drop" model of the nucleus.

THE ITALIAN NAVIGATOR—FERMI

The discovery of the neutron was seen immediately as a tremendous opportunity. One of the standard techniques in particle physics was projecting particles at various atoms. But there was a problem. It was difficult to get near the nucleus because it was positively charged and would repel positively charged particles. Furthermore, negatively charged particles would be repelled by the cloud of electrons in the outer part of the atom before they even got close to the nucleus. But neutrons had no charge and were therefore the ideal particle for bombardment of the nucleus. One of the first to see the potential was Enrico Fermi of Rome.

Born in Rome in 1901, Fermi was a sickly child and was cared for by nurses during much of his early youth.[5] He was particularly close to his older brother, and when he died at fifteen, Enrico was devastated. In his sorrow he buried himself in physics and mathematics books and soon became fascinated with science. He received his doctorate in 1922 just a few months before Mussolini seized power in Italy. His thesis, which was on x rays, was so well done and so complete that the members of his committee passed him quickly. It was obvious he knew everything about the area, and they couldn't think of any questions to ask him. Furthermore, it was clear to them that he would become a first-rate scientist. He did postgraduate work in Germany under Bohr, then returned to the University of Rome.

Fermi saw the potential of neutrons immediately and began an extensive program of bombarding all the elements of the periodic table with the particles. He had almost completed the program in 1934 when he discovered that he could slow neutrons down by passing them through paraffin, and these slow neutrons were even more effective than fast neutrons. In particular, when they penetrated the nucleus, they frequently created a new isotope. Most of these isotopes, however, were unstable and soon decayed.

When he bombarded uranium with neutrons, he got some particularly strange results and didn't know how to explain them. A German chemist suggested that he chemically analyze the products of the bombardment. But Fermi was too busy with his program, and he therefore missed one of the greatest discoveries of all time.

As Fermi continued his research, Europe was in turmoil. Hitler had taken over Germany, and Mussolini had allied himself with him. At the suggestion of Hitler, Mussolini issued a proclamation against the Jews. Their jobs, businesses, and passports were to be taken away. Fermi was not Jewish, but his wife, Laura, and their two children were. Fermi knew it was only a matter of time before they would be rounded up. He immediately wrote to four universities in the United States, and within months Fermi had five offers of employment.

The next problem was getting out of the country. He had heard he was in line for the 1938 Nobel Prize, but he knew it was still a long shot. If he won, he would have to go to Stockholm to collect the prize, and he could take his family with him. Day after day he waited for the news, but heard nothing. Then finally early one morning the phone rang: he had won. Once out of the country he would not return, but he knew he had to be careful. They wouldn't be able to take very much with them—it would alert the authorities—so they took only what they needed. The money from the Nobel Prize would tide them over for a while. After accepting the prize, the Fermis headed for New York and Columbia University.[6]

THE DISCOVERY OF FISSION

Fermi had not chemically analyzed the products of his bombardment of uranium, but others soon became interested in doing it. One of them was Otto Hahn, who worked at the Kaiser Wilhelm Institute in Berlin. Hahn had just read about an experiment performed by the Joliot-Curies in Paris in which they had bombarded uranium with neutrons. He was sure their interpretation of the experimental results was wrong, so he decided to redo the experiment and analyze the results for himself. Working with Hahn was Lise Meitner and Fritz Strassman. Meitner was born in Vienna in 1879 and had obtained her doctorate from the University of Vienna in 1906. Baptized in infancy, she was raised as a Protestant even though she was of Jewish descent. Just as she, Hahn, and Strassman began their historic work, Hitler issued a proclamation against Jews. Meitner hardly thought of herself as a Jew. She had not practiced the faith, but she knew she was of Jewish descent and that's all that mattered to the Nazis. She knew she had to leave Germany as quickly as possible, but she didn't have a valid visa. All she had was an expired Austrian visa, and it wouldn't be of much good to her. She also knew it was too late to get a new one because they would see immediately that she was Jewish and wouldn't let her out of the country. She got in touch with Bohr in Copenhagen, who made arrangements for her to enter Holland without a visa. But she still had to get through the Nazi patrol at the border. They would wonder how and why she got such permission. Furthermore, a proclamation had now been issued that no "academics" were allowed to leave the country.[7]

She took the train to the Dutch border. Writing later of the experience, she said, "At the border I got the scare of my life when a Nazi military patrol of five men going through the coaches picked up my Austrian passport, which had expired many years before. I got so scared my heart almost stopped beating." For ten minutes she sat as the Nazis looked over her passport and discussed it back and forth. Finally, they returned it to her without a word. Minutes later, to her relief, she was in Holland.

Bohr arranged a position for her in Stockholm, but she soon became depressed. Hahn was in the midst of one of the most important experiments he had ever done, and she had no money to do any research. In addition, she was in a foreign land, didn't know the language, and knew no one.

Meanwhile in Berlin, Hahn and Strassman had begun chemically analyzing the products of uranium bombardment. They soon discovered what they thought were two isotopes of radium. To Hahn this made no sense; radium was lighter than uranium. When a neutron struck a nucleus, it always produced heavier elements—not lighter ones. This confused him, and he knew he had to look into it further. Then just before the Christmas holidays of 1939 he got the surprise of his life. Among the products was the much lighter element barium. How could it have gotten there? He repeated the experiment, but there was no doubt. In a state of confusion, he wrote Meitner; he was sure she would know what was going on. As he waited for her reply, he began to write the first draft of a paper for publication. If Meitner could come up with some sort of an explanation, he would add it and put her name on the paper. Several days passed, then a week, and he heard nothing. Finally, he got a letter, but he was disappointed. She congratulated him on the discovery but made no attempt at an explanation.

Meitner was almost in shock when she received Hahn's letter. It seemed impossible that barium had been produced, but she had no explanation for it. It was close to Christmas, and she was lonely. She got in touch with her nephew, Otto Frisch, who was working in Copenhagen. He was single, and they had spent Christmas together before. Would he like to spend Christmas with her? He replied that he would, and they agreed to meet in the Swedish town of Kungälv.[8]

Frisch, a nuclear physicist, had been working on the magnetic properties of the nucleus. He was sure this was a good opportunity to get Meitner interested in his research. But she was not eager to hear about it; she handed him Hahn's letter almost as soon as they met. He read it and shook his head. Something was wrong, he said. Hahn had to be wrong; there was no way uranium could produce barium.

"Hahn does not make mistakes," Meitner said emphatically.

Frisch was confused. There didn't seem to be any alternative.

They decided to go for a walk to think things over. Snow covered the ground and Frisch had brought his skis, so he donned them while Meitner walked. They began discussing Gamow's liquid drop model of the nucleus; it had recently been pushed by Bohr. If the uranium nucleus was, indeed, like a drop of water, was it possible that it could break in half? If it absorbed a neutron, the energy might start it oscillating, and if it oscillated enough, it could take on the shape of a dumbbell. If this happened, the two ends of the drop would repel one another electrostatically, and it was possible that a splitting could occur.

They sat down on an old tree trunk as Meitner pulled out a piece of paper and a pencil.[9] What would the energy associated with the splitting of the water drop nucleus be? There would obviously be a strong repulsion just before it split. Meitner made the calculation. It would be approximately 200 million electron volts (an electron volt is the energy an electron gains in passing through a voltage difference of one volt)—not a large amount of energy, but if each atom gave off that amount when it split, it would be a large amount.

Then Meitner remembered Einstein's formula relating energy and mass. She added up the masses of the two product nuclei and compared it to the mass of uranium. It was less. She converted the mass difference to energy using Einstein's formula. It was 200 million electron volts—the same as she had got in the earlier calculation. This was no coincidence.

They had to convey their discovery to Bohr as quickly as possible.

There now seemed to be no doubt. When the uranium nucleus absorbed a neutron, it became unstable and fissioned, or broke in half, with the release of considerable energy (fig. 34). Frisch named the process *fission*. As soon as he got back to Copenhagen, he went directly to Bohr. He had barely begun to explain the process

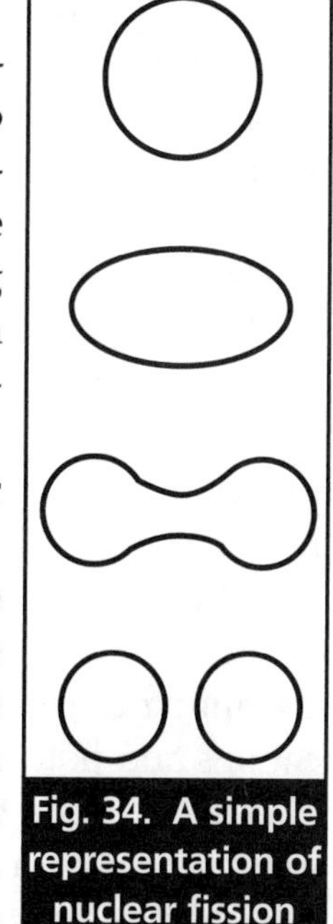

Fig. 34. A simple representation of nuclear fission

when Bohr struck his forehead with his hand. "Oh what idiots we all have been!" he said. "This is wonderful. This is as it must be." He asked Frisch if he had written up a paper for publication. Frisch said he had not. Bohr was leaving for the United States the next day and promised to keep it secret until he and Meitner could get it published.

Bohr spent the trip to America trying to understand the details of the process. He was accompanied by Leon Rosenfeld, and together they looked into some of the consequences of the strange new phenomenon. Enrico and Laura Fermi and John Wheeler met the boat in New York. Laura Fermi later said that she was aghast at how much Bohr had aged and how strained he looked. She knew, however, that for months he had been seriously worried about the situation in Europe and had helped numerous Jewish scientists escape from Germany. He had, in fact, put his own life in danger many times, but she didn't know that he was now also worried about something else. It was immediately obvious to Bohr that a bomb could be built, and the Germans would soon learn about the possibilities of fission and take advantage of them.

Bohr had promised to keep the breakthrough secret until it was published. But he forgot to tell Rosenfeld, who broke the news soon after they landed. When Fermi heard of Frisch and Meitner's work, he immediately thought of the strange results he had got in his experiments on uranium years before. Everything now made sense.

Bohr was shocked when he heard that everyone knew about the discovery, but there was little he could do about it. He therefore made an announcement at a Washington conference on theoretical physics that he was attending. Everyone was stunned at the news. Many of the world's top physicists were at the meeting: Harold Urey, Otto Stern, Gregory Breit, Isaac Rabi, George Gamow, Edward Teller, Hans Bethe, and others. After Bohr made the announcement, Fermi took the stand and explained the implications and potential of the discovery.

One of those in the audience was Leo Szilard, who had come to America from Germany several years earlier.[10] The son of a Jewish engineer, he had fled Germany shortly after Hitler came to power. His first stop was London, and it was here in 1934 that he had conceived the idea that the nucleus might split, and if it did, considerable energy would be released. He had no idea how it would take place, but he understood the implications. According to a well-known story, he was waiting for a red light on a London street when it suddenly dawned on him that energetic neutrons would be released if the nucleus did split, and these neutrons would cause further nuclei to split. A "chain reaction" would occur with all, or most, of the nuclei in the sample splitting, and a tremendous amount of energy would be released. Over the next few weeks, he applied for a patent on the process, but he kept it secret, fearing its potential for a bomb.

Leo Szilard

When Szilard heard of the discovery of fission, he knew the likely consequences. He worried about Germany building a bomb and immediately talked to Fermi about the possibility. Fermi didn't take him seriously, but he decided to speak to the military anyway. An appointment was arranged for him to talk to U.S. Navy officials in Washington, D.C.

As Fermi sat in the lobby waiting for the meeting, someone opened the door and yelled back at the officer that Fermi was to meet: "There's a 'wop' out here waiting for you." (He obviously didn't know the "wop" had just won the Nobel Prize and would literally change the world in a few months.) The meeting was not a success. Several high-ranking officers listened politely as Fermi explained fission and the possibility that the Germans might build a bomb using it. They told him they would get in touch with him, but Fermi knew there was little chance of that.[11]

Szilard was particularly disappointed at the results of Fermi's meeting. He was now becoming concerned about the large reserves of uranium in the Belgian Congo and was sure the Germans would eventually go after them. He had to warn Belgium not to sell uranium to Germany. But how? Then he thought of Einstein; he had worked with him earlier, and knew that Einstein was a friend of Queen Elizabeth of Belgium. Szilard phoned the Princeton Institute of Advanced Study, where Einstein had his office, and was told that Einstein was at his summer home on Long Island. He phoned and made an appointment with him.

Szilard could not drive a car, so he had to get Isaac Rabi to drive him. They had trouble finding Einstein's home and were almost ready to give up and go back when they spotted a young boy sitting by the curb. Einstein was so well known everyone would have to know where he lived. They asked the boy and soon had instructions. Einstein greeted them and invited them in. He had heard nothing about fission and chain reactions but immediately realized the potential for a bomb. He didn't want to bother Queen Elizabeth but agreed to forward a letter to a friend in the Belgian cabinet. Szilard also talked about sending a letter to the White House, but it would have to be personally delivered if it was to have any impact.[12]

Enrico Fermi

Einstein agreed to sign the letter. On the trip back to New York, Szilard began to think about who he could get to deliver the letter to President Franklin Roosevelt. It had to be someone who had access to him. He remembered Alexander Sachs, a vice president of the Lehman corporation and a biologist. As soon as he got back, he got in touch with Sachs and arranged a meeting with him. Sachs took an immediate interest and agreed to do everything he could. Szilard then composed a letter and made another appointment with Einstein. This time Edward Teller drove him. Einstein read the letter, made a few small changes, and signed it. Szilard passed the letter to Sachs on August 15, 1939.

Sachs tried to get an appointment with President Roosevelt but was put off. The president was busy with the war effort and had little time for visitors. Szilard waited patiently, but after a month he had still heard nothing. Finally he went to Sachs and was disappointed to find that the letter was still on Sachs's desk. Sachs tried again, and a few weeks later finally got an audience with President Roosevelt.

"I'm going to tell you something very important," Sachs began when he finally gained access to the president. He gave the letter to him and explained what he knew about the German threat. President Roosevelt assured him that action would be taken. Almost immediately he set up a committee to look into the possibilities; it was referred to as the *uranium committee.*

Things couldn't happen fast enough for Szilard. He expected immediate action and fretted as the days passed with nothing happening. Finally a few weeks later $6,000 was allotted for research at Columbia University. But it was soon discovered that there was a problem. The first thing that was needed was a "slowed-down" version of the bomb—a *nuclear reactor*, and for this it was necessary to have something that controlled the neutrons coming out of the fission process. This "something" is now referred to as a *moderator*, and it was known that two materials were excellent modulators: heavy water and graphite. Heavy water was very expensive, so graphite was the better choice. When Fermi made an estimate of the size of the "pile" of graphite bricks that would be needed, he found it was too large to be housed in any building at Columbia University. They would have to move to a larger facility.

Fermi was not happy. He had bought a house in the suburbs of

New York and had settled in. But there seemed to be no alternative. A search of nearby institutions was made, and the best bet appeared to be the University of Chicago. Arthur Compton, the department head, offered his facilities, but the only place high enough for the pile was the racket courts under the stands at the football stadium—Stagg Field. And it was unheated.

THE FIRST NUCLEAR REACTOR

Construction of the first nuclear reactor began in Chicago on November 16, 1942. A nuclear reactor is a slowed-down version of the atomic bomb.[13] It was needed to see if the bomb could be built, and how to design it. Fermi was in charge of the project. The pile was to consist of seventy-six layers of graphite bricks, with each being 4 inches by 4 inches by 12 inches. The overall shape of the pile was to be a sphere. It was made up by piling one layer of pure graphite bricks, then two layers with the bricks drilled and loaded with uranium. As the structure increased in size, a lumber scaffolding had to be constructed to get to the upper layers.

Fermi organized his team into two groups, one under Walter Zinn and the other under Herbert Anderson. They each worked twelve-hour shifts. Cadmium rods were placed within the graphite as a further control on the number of neutrons released. They would allow Fermi to approach the *critical size* (the size when it became critical; in other words, the reaction became self-sustaining) carefully. Neutron counters were also placed within the pile so that the progress of the reaction could be monitored.

The critical factor was referred to as k. It was a measure of the number of neutrons that were being generated in the reaction. When k became greater than one, the pile was critical and the nuclear reaction was self-sustaining. Fermi wanted to be sure it barely went above 1.0, and therefore could be controlled. If it accidentally went too far above 1.0, the reaction could get out of control.

By late November the pile was nearing completion. The weather was bitterly cold. As each layer was added, k got closer and closer to 1.0. On the evening of December 1, Zinn noted that the rate of neutron production increased rapidly and the reaction would likely go critical with one more layer of bricks. He phoned Fermi and was told that he had already gone to bed for the evening. December 2, 1942, was therefore the fateful day.

Fermi made his way to the makeshift lab early the next morning. There were several inches of snow on the ground and the temperature was below zero. The last layer of bricks was added. The only thing stopping the pile from going critical now was the cadmium rods. When they were pulled out, it would go critical.

A crowd formed on the balcony overlooking the pile. Almost everyone involved in the project was there. The only noise in the large room was the clicking of the neutron counters. Fermi made some calculations using his tiny, six-inch slide rule, then ordered his assistant George Weil to pull the cadmium rod out six inches. The clicking of the counters increased. He made some more calculations and ordered the rod to be pulled out further.

The clicks increased again, and it was obvious they were close to criticality. Everyone was tense with anticipation. Suddenly there

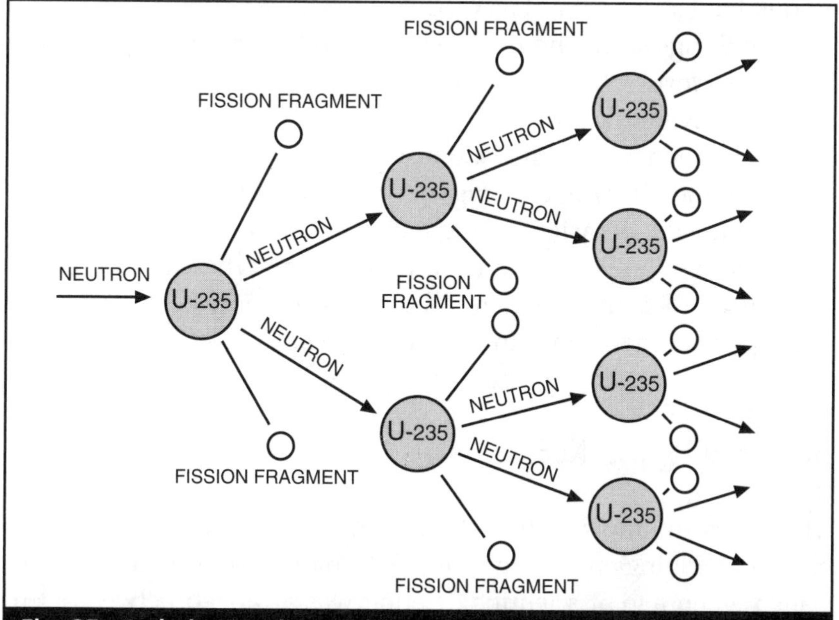

Fig. 35. A chain reaction with neutrons triggering further fissions

was a loud crack and everyone jumped. One of the safety rods had released. It had been set to go off just before *k* hit 1.0.

It was 11:30. Fermi sighed. "Let's go to lunch," he said.

After lunch they assembled again, and Weil set the cadmium rod to where it had been just before lunch. The counters were clicking rapidly. Fermi ordered Weil to pull the rod out a few inches more.

Suddenly the counters went wild. They became a roar. Everyone realized what had happened. After a few moments Fermi raised his hand, "The pile has gone critical."

Everyone expected him to close it down immediately. If it continued too long, it would melt down and everyone in the room would be killed. Several minutes passed. He announced that *k* was 1.0006. Everyone waited in anticipation. How long would he let it go? Finally, after four and a half minutes, he ordered Weil to push in the cadmium rod and shut it down.

The time was 3:25 P.M. Most scientists now regard 3:25 P.M., December 2, 1942, as the dawn of the atomic, or nuclear, age.

One of those in the balcony was Arthur Compton, the head of the physics department at the University of Chicago. He rushed to phone James Conant at Harvard University. Knowing that he had to keep things secret, he said, "The Italian Navigator has landed in the New World." It was a code.

"How were the natives?" asked Conant.

"Very friendly," replied Compton.[14]

Team member Eugene Wigner of Princeton had bought a bottle of wine for the occasion. Using paper cups, they all made a toast. In the crowd, however, there was one person who was not happy. He was Leo Szilard. He knew a bomb would now be built, and he was apprehensive about the consequences.

BOHR MEETS HEISENBERG

There was still much anticipation about how far the Germans were from developing an atomic bomb. Bohr had just returned to Copenhagen when a joint scientific meeting was arranged between Denmark and Germany in October 1941.[15] Bohr had no use for such meetings and usually boycotted them. But Heisenberg was coming to this meeting, and Bohr was anxious to talk to him. He was still worried about the possibility of a German atomic bomb.

The evening after Heisenberg arrived, the two men went for a walk. Heisenberg still looked upon Bohr as a father figure and greatly admired him. Soon after the walk began, Heisenberg asked Bohr about the morality of working on the "uranium project." He knew he had to be careful in the way he selected his words because there was no doubt that Bohr was being watched by the Germans. Bohr was taken back by the question and asked Heisenberg if he thought a bomb was really possible. Heisenberg said he thought it was, but it would require a tremendous technical effort, and he hoped it wouldn't happen.

Bohr was shocked. He was now sure the Germans knew everything. Heisenberg began telling him about the project he was working on, but Bohr hardly heard him. In an obvious breach of security, Heisenberg sketched the outline of a heavy-water reactor he was building. The Germans knew about graphite but had decided to use heavy water. Bohr took the diagram and later showed it to authorities.

When the two men parted, Bohr was still in shock, and the meeting affected him for months. Heisenberg also left the meeting in "a state of confusion and despair," according to an account his wife later wrote. He was sure he hadn't convinced Bohr of his true feelings about the morality of the bomb.

OPPENHEIMER AND LOS ALAMOS

Once it was proven that a bomb was possible, the next stage was building it, and everyone involved knew it would be a massive project. Brig. Gen. L. R. Groves was selected as the commanding officer in charge of the overall project. One of the first problems was the separation of enough uranium 235 from natural uranium (natural uranium is a mixture of uranium 235 and uranium 238)

Robert Oppenheimer

for the bomb. A huge plant was set up for this in Oak Ridge, Tennessee. In addition, a central laboratory was needed where the problems related to the construction and engineering of the bomb could be ironed out. A director would be needed for the facility. Robert Oppenheimer was soon seen as a potential candidate.

Oppenheimer was born in New York City in 1904. He graduated from Harvard University in 1925 and went to England to work under Rutherford. To his disappointment he was assigned an experimental project under G. P. Thomson. He disliked the work and was considering quitting. Then one day Bohr visited the lab and talked with him. Oppenheimer was so impressed with him he made up his mind immediately to become a theoretical physicist. He left Cambridge for Göttingen, Germany, and received his Ph.D. from the University of Göttingen a few years later. Quantum mechanics was all the rage at this time, and Oppenheimer soon mastered the new theory. Working with Max Born, he developed a theory for explaining the collision of particles.

In 1928 Oppenheimer returned to the United States. Several offers of employment had been made to him, and he had finally accepted a joint appointment at the University of California and Caltech, also in California. He was one of very few who was familiar with quantum mechanics, and students soon began to flock to his classes to learn about the new developments. Over the next few years, he and his students published several important papers—one predicting the existence of neutron stars and another predicting black holes.

Oppenheimer knew about the success at the University of Chicago and was anxious to contribute to the project. He met with General Groves several times, stressing the importance of a central lab. Groves was impressed with him and began to think of him as a possible director. "He's a genius," Groves said. "He knows . . . everything . . . except sports."[16]

When Hans Bethe heard that Oppenheimer was being considered as director, he couldn't believe it. Oppenheimer had no experience in directing people, and he was known to be quite abrasive and sarcastic at times. Furthermore, he had no Nobel Prize, and he would be directing dozens of people with Nobel Prizes. One of the most serious problems, however, was the rumor of his left-wing activities. But there was no doubt that he was an excellent nuclear physicist and was very capable.

Groves submitted Oppenheimer's name to the military policy committee. They were not happy and suggested that he come up with another name. After looking into it further, Groves became even more convinced that Oppenheimer was the right man. He went back to the committee and submitted Oppenheimer's name again. They objected, but Groves persisted and finally convinced them he was the best man.

The next problem was the location of the new facility. It had to be kept secret and therefore had to be isolated, but it couldn't be too isolated. Several sites were considered, and they finally narrowed in on Jemez Springs, New Mexico, which was about forty miles northwest of Santa Fe. The site was named Los Alamos, after one of the nearby canyons. With the site for the new lab selected, Oppenheimer immediately began crisscrossing the country looking for the best people to work there. Without giving away the details, he managed to convince most of the top physicists in the country of the importance of the project.

Oppenheimer moved to Santa Fe on March 15, 1943. The lab was not yet ready, but over the next few weeks it was completed.[17] It consisted of several one-story, barrack-like buildings and a few two-story apartment buildings. In April Oppenheimer assembled the scientists, and he and his former student, Robert Serber, told them that the object of the project was to build the world's first *atomic bomb*.

Several problems had to be overcome. One of the first was the critical size of the uranium 235, in other words the size of the *core* that was needed for the bomb. Too small and it would not fission properly; too large and the initial reactions would blow the material apart too rapidly. Another problem was the detonation of the bomb. The uranium 235 had to be kept separated in subcritical pieces, then brought together powerfully to the critical size for the explosion to occur. Several different designs were considered.

Over the next couple of years, the problems were resolved, and on July 16, 1945, the bomb was ready for its first test. The Alamogordo Air Base was selected for the test; it was an isolated region of mountain-rimmed desert. The city of Alamogordo was 50 miles away. The bomb tower was referred to as *ground zero*. Three observation points were set up at distances of 5.7 miles from ground zero. Other observation points were set up further back.

The test was scheduled for 4:00 A.M., but rain and lightning forced a delay until 5:30 A.M. Everyone who had worked on the bomb and a large group of military men made up most of the observers.

The countdown began about twenty minutes before 5:30 A.M. Tension began to mount as the minutes were called out. At "Zero minus ten seconds," a rocket rose into the air and gave off a flare. Then at "Zero minus three seconds," another one rose.

Suddenly there was a blinding flash, then a roar, followed by a pressure wave that knocked several men off their feet. From the position of the tower, a huge multicolored cloud rose into the sky. The first atomic blast had taken place.[18]

The atomic bomb was a reality.

A few years later in November 1952, an even more powerful bomb—the *hydrogen bomb*—was exploded in the South Pacific.

THE SPIN-OFFS OF NUCLEAR ENERGY

Our ability to build the atomic bomb depended on our understanding of the nucleus and nuclear energy, which in turn depends strongly on quantum mechanics. The basic model of the nucleus is a quantum mechanical *well* in which the protons and neutrons are trapped. This nuclear well can be viewed as resulting from the mutual attraction of all the protons and neutrons (usually referred to as *nucleons*) over the dimensions of the nucleus, which is about 10^{-12} cm. Schrödinger's quantum mechanical equation is used to calculate the energy levels within the well, and the nucleons reside in these levels in much the same way electrons reside in various energy levels (orbits) around the nucleus. Each of the nucleons has its own particular quantum numbers (numbers specifying its exact state).

Radioactivity, as we saw earlier, is explained by "quantum tunneling" out of the well. A given nucleon on an energy level within the well hits the barrier repeatedly. According to classical theory, there is no possibility of it penetrating the barrier, but quantum mechanics tells us that there is a small probability of it getting through. And indeed we see a small number getting through in the form of radioactive decay.

Quantum mechanics is also important in explaining the force between the nucleons. For many years the accepted theory was the meson theory in which it is assumed that mesons passing back and forth between the nucleons generate the force. In recent years this theory has been superseded by quark theory in which the nucleons are made up of quarks and the particles that are passed back and forth are gluons. Quantum mechanics is important in both theories. So, in general, quantum mechanics plays a large role in our understanding of the nucleus and nuclear energy.

The benefits to society of nuclear physics are numerous. Much of the energy that is used to generate electricity in the United States and in other countries of the world comes from nuclear reactors. The first submarine driven by nuclear power, the *Nautilus*, was launched in 1954. The first nuclear batteries, called SNAP (system for nuclear auxiliary power), were built in 1956. Then in

1957 a nuclear reactor in California produced the first electric power for civilian use. It was followed by nuclear power stations in Britain and the Soviet Union. Then in 1959 the NS *Savannah*, a merchant ship driven by a nuclear reactor, was launched. And in 1961 nuclear energy was used for the first time in a space satellite.

Although the link to quantum mechanics is less direct, the extensive use of radioisotopes, radiochemicals, and so on, in medicine has been a tremendous breakthrough. It has led to a branch of medicine known as *nuclear medicine*. Some of the uses of nuclear materials are in early detection of cancer; treatment of cancer; brain imaging; and thyroid, lung, and gastrointestinal studies.

Radioisotopes are also now used extensively in agriculture and food technology. They have helped increase food production and have extended the life of perishable foods. Besides this they have found many practical applications in devices such as smoke detectors and pacemakers. All in all, they have made a huge contribution to society.

The Computer Revolution

It might not seem that quantum mechanics had much to do with the computer revolution, but it did. Early computers were purely mechanical, so quantum mechanics played no role in their development. But once transistors and microcircuits were used, quantum mechanics did play a vital role. And indeed it will no doubt play an increasingly important role in the computers of the future as the components get closer and closer to the quantum realm.

Even though early computers did not depend on quantum mechanics in any way, I will include a discussion of them so that the story of computers is complete. The first calculating machines were built by Blaise Pascal of France and, independently, Gottfried Leibnitz of Germany, but it was Charles Babbage of England who is generally thought of as the grandfather of computers. There's no doubt that he was years ahead of his time—perhaps as much as a hundred years—and it created many problems for him. Society had little need for computers in the mid-1800s and showed little

Charles Babbage

interest in them. Babbage had so many failures and so many problems in his struggle to build a computer he eventually became a cantankerous, angry-spirited old man.

Born in London in 1791, Babbage came from a well-to-do family. He showed an early interest in mechanical things and was always asking how things worked. To the dismay of his parents, if he wasn't told, the object was soon taken apart.[1]

Babbage, who attended a private boarding school in his youth, developed an early interest in mathematics. He would frequently sneak out at night and study mathematics books by candlelight. By the time he went to Cambridge, he already knew more mathematics than most of his teachers and spent much of his time studying papers and texts that were beyond the class. In collaboration with John Herschel (son of astronomer William Herschel), he set up the "analytic society." The aim of the society was to keep abreast of the latest developments in mathematics. One of its projects was the translation of a well-known text on calculus. Working with Herschel, Babbage not only translated it into English but added considerable text of his own. It eventually became the standard calculus text in England.

Babbage led an active social life while at Cambridge. He joined many groups and had several interests outside of mathematics: optics, weather, geology, and ghosts were just a few of them. He graduated in 1814, expecting to get a job at one of the colleges, but after an extensive search he found nothing. He had few connections in the right places, which at that time was necessary. He finally gave up and moved to London; he now had an inheritance from his father, so he was under no financial pressure. He continued his research in London.

Babbage was interested in many different areas, but he had a passion for numerical tables, in particular astronomical and mathematical tables, and he soon became concerned with the large number of errors that existed in the published tables. He eventually produced and published the most accurate logarithmic tables that had ever been formulated.

While at Cambridge, he had thought about the possibility of "machines" for calculating the numbers in the various types of tables. One evening after a long, frustrating day working at the analytic lab he put his head down on the table and fell asleep. A colleague came into the room. "Babbage, what are you dreaming about?" he yelled at him.[2]

Babbage shook his head as he woke up. "I was dreaming that I had invented a machine to calculate all these numbers," he said, pointing at the sheets covered with calculations that lay in front of him.

Over the following months and years, he continued to think about the idea and was soon formulating a plan. By the early 1820s he knew generally how he would proceed, so he wrote to Sir Humphrey Davies, the president of the Royal Society, describing his ideas and telling him of the importance of such a machine.

Davies was interested and passed the letter on to members of the British government. A committee was set up to consider it, and in May 1823 Babbage was awarded a grant of $7,500 for the project. He knew it wouldn't be enough, but it would get him started, and he had enough of his own money to continue if funds ran out. He was sure the project would take only two or three years. Although he knew there were still many problems to be resolved, he had already worked out the basics of the design. The machine would be entirely mechanical and would consist of thousands of tiny gears and linkages, many of them quite intricate.

He soon found that no one could make the gears and linkages that he needed. But Babbage didn't let this stop him; he rolled up his sleeves and made his own tools for producing the necessary gears and linkages. In the process he made tremendous strides in toolmaking, but it delayed the construction of his machine, putting considerable pressure on him. His health soon began to deteriorate, and in October 1827 he had a breakdown. The doctor told him he had to take a rest. Babbage was disappointed but went to Italy to recuperate.

Upon his return to England, Babbage began working on the project again but soon ran into problems. He had used up the money from his grant and a considerable amount of his own money, and when he tried to get more, he was rejected. Over the next few years, he had so many difficulties it began to affect not only his health but his outlook on life. He became bitter, depressed, and resentful.

Actually, he brought many of the problems upon himself. For example, he became fed up with the Royal Society and the unfavorable climate for research and science in England, and he wrote a scathing book that advocated new reforms—but it did little good. Furthermore, he took some of his anger out on the street musicians and organ grinders of London by writing letters to papers against them, but it also backfired. The musicians began following him around, badgering him everywhere he went. Some of them even set up camp around his house and played music all night, much to his annoyance. He also wrote scathing letters against the prostitutes of London, and they also set up camp around his house as a further annoyance. At one point more than a hundred people followed him, hurling obscenities at him as he went for a walk. It's perhaps little wonder that he had such a poor outlook on life.

Babbage's first computer was referred to as his *difference machine* because it was based on a mathematical tool referred to as the method of differences. Well before it was complete (actually, he never did complete it), he began working on the design of a machine that would calculate analytic functions. He referred to it as his *analytic machine*. It was amazingly close to the concept of the modern computer; it even had a simple memory.

Babbage continued to fight with government officials over

money. In most cases he was attempting to get back money he had already spent. After he completed the design for his analytic machine, he approached the government to help pay for it. He had already spent a considerable sum of public money on his difference machine and had not completed it, so they were not happy. Nevertheless, they decided to call a meeting to consider it; after months of haggling, they couldn't come to a decision, and while they delayed, the government changed. So Babbage had to start all over again.

He approached the new government, but they were no better. They procrastinated for several years, until Babbage finally sent a letter to the prime minister in an attempt to get a decision. A meeting was called, and a decision was made: all government support to his projects was cut off.

Babbage was disappointed. He had gone through so much of his own money that he now had little left. Over the next few years, he continued trying to get money, but he never did finish either of his machines. Nevertheless, he did make important advances and aptly deserves the title "grandfather of computers." As Isaac Asimov writes in his biography of Babbage, "Most of his life was spent in a vast failure that seemed a success only in hindsight."[3] An irony of the situation is that shortly after the British government turned Babbage down, it decided to support a similar project in Sweden.

THE FIRST "DIFFERENCE MACHINE"

After Babbage died few advances were made. No one needed a computer and there was little interest, so it was many years before further progress was made.

The first *difference machine* that was actually built was one at MIT in the 1930s. Vannevar Bush was attempting to solve some differential equations associated with an electrical network he was working on when he decided that it could be done much better with a machine, so he began designing one.

Born in Massachusetts in 1890, Bush graduated from Tufts University. He received his doctorate from MIT and Harvard. He and several colleagues constructed a machine that consisted of three basic units: an integrator, a multiplier, and a unit for addition and subtraction. It was completed in the 1930s. Although it was hard to set up, once it was going it could crank out a lot of data. Several copies of the device were built before World War II, with some of them being used for calculating ballistic trajectories.

THE Z MACHINES

The scene then changed to Germany and a civil engineer by the name of Konrad Zuse.[4] Born in Berlin in 1910, Zuse graduated from Technische Hochschule of Berlin in the early 1930s. We usually think of civil engineers as building roads and bridges, but Zuse went to work in an airplane factory designing airplane engines. He had a flair for mathematics, but even as a student he hated the tedious numerical calculations that were required in many engineering problems. The situation seemed to be no different in the design of airplanes, so he decided to build a calculator to do the tedious calculations he kept encountering. Strangely, he knew nothing about Babbage's or anyone else's work on computers and therefore had to figure everything out himself. He decided immediately that the machine should be based on *binary numbers*. Binary numbers were particularly adaptable to machines in that they only had two digits, zero and one, and any number could be made up of a combination of them.

The only space Zuse had available to him was his parents' living room in Berlin, so, to their annoyance, he set up a lab there and began working on his *calculator* in his spare time. By 1936 the device was complete enough to apply for a patent. The following year he finished it, calling it Z1. It was a simple system capable of storing and manipulating several numbers.

He saw immediately, however, from what he had learned that he could build a much better model, and he was soon constructing

Z2. Indeed, he soon got so wrapped up in Z2 that he quit his job and began working on it full time. He even recruited a friend, a graduate student, to join the project. Zuse's Z2 was almost finished when Germany went to war in 1939; he was called up, and his project was put on hold.

He was lucky enough, however, to have several good friends. They petitioned the government to have him released from the army so he could finished the Z2, which they emphasized might be important to the war effort. After considerable paperwork Zuse was finally released, but he had to go back to the airplane factory and could work on his computer only in his spare time. Finally, in late 1939 he finished it.

Even before it was finished, he had begun designing an improved model. As with the two previous models, it was based on relay technology (a relay is a simple device which may be mechanical or electronic and which has two states: off and on. It can be used to represent the two numbers of the binary number system, namely 0 and 1). With the war, however, parts were becoming harder and harder to get. Zuse's student friend began encouraging him to use more electronic parts, but Zuse decided to stick with the mechanical relays. Money was always a problem, but Zuse finally got the support of the German Aeronautical Research Institute. Strangely, even with money from there, he still continued to work in his parents' living room.

The Z3, his latest machine, was considerably faster than the Z2. It had a sixty-four-word memory unit and could perform approximately four additions per second. But before he could finish the Z3, he was drafted again, and this time he was sent to the eastern front. Again his friends came to his rescue. They went to the German Aeronautical Research Institute and encouraged them to try to get him released, and a few weeks later he was back in Berlin. Soon he had completed the Z3 and was designing an even more advanced model, the Z4.

By now Berlin was being heavily bombed and there was considerable danger that the Z4 would be destroyed, so Zuse moved it to Göttingen. But Göttingen was soon surrounded and in danger of falling to the enemy, so he took it to a house in Bavaria and hid it in the basement. Then he, along with rocket expert Werner von Braun, went into hiding. Finally, however, they gave themselves up to the Americans, and in 1948 Zuse was taken to London for interrogation. He told authorities about his computer, but they didn't know what he was talking about (part of the problem was a language barrier) and never took him seriously. Upon his return to Germany, he retrieved the Z4 from Bavaria and completed it. It was eventually taken to ETH in Switzerland where it was used for several years. Zuse formed a company and continued designing and manufacturing computers. He eventually became interested in computer languages and software and made several important advances.

STIBITZ AND HIS RELAYS

Zuse was not the only one using electromagnetic relays. At Bell Labs in the United States, relays had been used for years in telephone switching circuits. In 1937 George Stibitz, a mathematician working at Bell Labs, took several of these relays out of the junk pile one day and began playing with them. They had two states—on or off—and were similar to binary numbers. A binary number could therefore be stored in a set of relays.

He took several of them home, and using an old can and some flashlight batteries, he built a simple device that could add two binary numbers. Pleased with his accomplishment, he brought it to work to show his colleagues. Several months later he was approached by his boss and asked if he could modify the device so it would work with *complex numbers*. (A complex number has two parts: a real part and an imaginary part. The real part is a number such as 1, 2, 3, . . . ; the imaginary part is an imaginary number such as the square root of one.) Stibitz was sure he could, and over the next few weeks, he made a model that was able to manipulate complex numbers. Bell then decided to follow up on the device, and over the next couple of years, they built a larger, more powerful device. In January 1940 it was ready. It did not have the capability of storing programs, but it could perform mathematical tasks at high speeds, which made it invaluable for doing ballistic calculations during the war. After the war several more advanced models were built.

MARK I AND ABC

About the time Stibitz and Zuse were working on their machines, a third effort was underway at Harvard University under the direction of Howard Aiken. Aiken was a professor of applied mathematics in engineering at Harvard. He was interested in a particular type of differential equation referred to as *nonlinear* (an equation is nonlinear when two solutions to it do not give a valid solution when added together). Nonlinear equations were very difficult to solve, and about the only way to do it was to solve them numerically. This meant a lot of number grinding. Aiken decided to build a machine that would do the calculations, and he began searching for somebody to support his project. Finally, after a long search, IBM expressed an interest and gave him a contract.

The device, which was eventually called Mark I, was completed in January 1943. It was considerably larger than any computer that had been built up to this time. It was fifty-one feet long and stood eight feet high. Inside the cabinets were 750,000 parts, which included switches, relays, regulators of various types, and wheels. When it was completed, IBM donated it to Harvard. Set up at Harvard in May 1944, it was used in the war effort. After the war Aiken went on to build Mark II for the U.S. Navy. He was also involved with two later models, Mark III and Mark IV.

About this same time, a different type of computer was being developed at Iowa State University. Mark I, like its predecessors, was based on relays and switches; the model built at Iowa State University, on the other hand, was based on electronic tubes. It was, indeed, the first electronic digital computer, and was referred to as ABC (the Atanasoff-Berry computer). Like so many others, John Atanasoff of Iowa State became annoyed at the large number of tedious calculations he was having to do in relation to a research problem. After a particularly frustrating day, he decided to take a long ride to relax. While he drove, he began thinking about how a calculating machine could be built. By the time he got back, he had the outline of the design in his head. He sketched it out, and soon a graduate student, Clifford Berry, took an interest in the project and began helping him. By December 1939 they had designed all the components and were ready to build.

Atanasoff got a small grant of $5,000, and they began to build the device in 1940. When finished, it contained 600 vacuum tubes—300 were in the arithmetic unit and 300 in various controls—and a simple memory unit. The memory used *capacitors* (charge-storing devices) mounted in two rotating drums. Unfortunately, the machine was not used extensively. About the time it was completed, Atanasoff left to join the war effort and went to the Naval Ordinance Lab in Washington. He left the machine behind.

ENIAC

Like the Atanasoff-Berry computer, another computer that was built about the same time also employed electronic tubes. Called ENIAC, it was the first large computer to be based on electronics and electronic tubes.[5] At the time it was a hundred times larger than any single piece of electronic equipment that had ever been built. Many important advances in computer technology were made during its construction.

Although many people worked on the machine, two people were instrumental in getting it started, and they were connected with it throughout its construction. They were J. Presper Eckert and John Mauchly.

Eckert was a student at the Moore School of Electrical Engineering at the University of Pennsylvania where the computer was built. He had been assigned to direct a student lab in 1941, and one of the students in the lab was John Mauchly, a professor from a nearby college. Mauchly had some training in electronics but thought he should brush up on the subject, so he enrolled in a course at the Moore School of Engineering. It soon became obvious that Mauchly was completely familiar with the material of the lab, so instead of performing the experiments, he sat on one of the tables and chitchatted with Eckert about electronics.

The talk soon turned to computers and how one might be designed. Shortly thereafter a position became available at Moore, and Mauchly was hired. Their plans then took on a much more serious nature, and Mauchly wrote a paper titled "The Use of High Speed Vacuum Tube Devices for Calculation." In the paper he compared electronic tubes to mechanical relays, showing that tubes would be much faster and more effective.

To build a computer they would, of course, need money, so they began searching for support but found no one that was interested. Then fate stepped in. It was 1943, and one of the major projects associated with the war was the calculation of ballistic tables for long-range gunners. The program had fallen seriously behind, and the army commissioned Lt. Herman Goldstine of the University of Michigan to find a way of speeding up the program. Goldstine heard of Eckert and Mauchly's work at Moore and went to visit them in March of that year.

Goldstine was impressed with their design and persuaded the military to support it. The computer was referred to as the "Electronic Numerical Integrator" or ENI. Later the words "and computer" were added, and it became ENIAC. It was soon obvious that the project would be a major undertaking and would involve a lot of people. Work began in May, with Eckert and Mauchly directing the project and overseeing the building of the computer.

Overall, the computer would consist of arithmetic units, control units, accumulators, and input and output devices. A card reader was used for input. By the time it was finished 18,000 vacuum tubes had been used, along with 70,000 resistors, 10,000 capacitors, and 1,500 relays; it was over a hundred feet long. Because of the large number of vacuum tubes and resistors, both of which generated heat, an extensive forced-air system was needed for cooling. With so many tubes, however, there was always the problem of one of them failing. Eckert designed the circuits carefully, however, so that a minimum of problems would develop. The device had no memory as we now know it, but it wasn't long before Eckert and Mauchly realized that this was a shortcoming and began to think about a more advanced version with stored-program capability.

ENIAC was the first really large electronic computer and for several years was the only such device in the world in daily use. According to some scientists, ENIAC performed more arithmetic calculations during its lifetime than had been done by all humans in the history of the world before 1945.

STORED PROGRAMS

ENIAC was not capable of storing programs for later use, but a version of it that had this capability was soon designed. Instrumental in much of the design was mathematician and physicist John von Neumann of the Princeton Institute of Advanced Study. Upon hearing about ENIAC, von Neumann visited the Moore School in 1944. He was quick to realize the importance of the machine and also saw some of its shortcomings. He soon became a frequent visitor to the Moore School and along with Eckert and Mauchly began to design a new machine that was named EDVAC. The major difference between ENIAC and EDVAC was that EDVAC would have program storage. This would turn out to be a tremendous advance in computer technology.

It is not known exactly how von Neumann contributed to the project, but with his tremendous insight and ability, no one would doubt that he made important contributions to it. The original concept, however, came from Eckert and Mauchly. Because of the contracts that Eckert and Mauchly were under, they were forced to keep everything secret. Von Neumann, on the other hand, was under no such restriction. Furthermore, von Neumann was well known for his ability to write comprehensive reports in which all known knowledge of a subject was brought to bear. After working with Eckert and Mauchly for some time, he wrote an extensive report titled "First Draft of a Report on EDVAC." It was not published, but was widely distributed, and it did not include Eckert's and Mauchly's names on it. Their being named as coauthors would have gone against the secrecy required of their contract. Nevertheless, they were annoyed.

Difficulties between von Neumann and Eckert and Mauchly began to compound as people began to refer to EDVAC as "Neumann's machine." To make matters worse, Eckert and Mauchly began having problems with the administration of the University of Pennsylvania. They had understood from the beginning that they had patent rights on ENIAC, but university officials were reluctant to sign over these rights. After several months of bickering, university officials demanded that Eckert and Mauchly sign over all patent rights to the university. They refused, and within a short time they left the Moore School. Shortly thereafter they formed a computer company. In the meantime von Neumann had also left the project.

THE TURING ENIGMA

Over the next few years, some of the most important advances in computer design were made in England. It might seem unlikely that Bletchley Park, outside London, England, would be a place for advances to be made. After all, it was a code-breaking facility. But significant advances were made there.

The story begins with the German invention of a message-coding device called Enigma. Developed in the 1930s, Enigma was first noticed by the Poles in the years before World War II. In concept, it was a simple device consisting of a typewriter keyboard with three wheels inside a box beneath it. There were electrical contacts on each side of the wheels, and each wheel had twenty-six contacts corresponding to the letters of the alphabet. When a message was typed in, it was sent to the second wheel via electrical contacts, but contact was made at a different position on the second wheel. Similarly, contact was then passed through to the

Alan Turing

third wheel, and again contacts were made with the wheels in different positions. With such a setup, 100 billion different codes could be produced, and, of most importance, the codes could be changed each time the machine was used. To interpret the codes, the receiver merely had to have his machine set in the same way as the sender. The system seemed foolproof.

The Poles began intercepting messages several years before the war and made some progress in deciphering them after they called in a well-known Polish mathematician. Other countries also intercepted the messages, but they made little progress in deciphering them. When the war began, Poland shared its knowledge with the Allies. Nevertheless, the machine and its code were such a serious problem that authorities set up a decoding facility to deal with it. To make things worse, the Germans made the machine even more complicated after the war began by adding more wheels to it.

One of the people hired at Bletchley Park was Alan Turing.[6] Born in London on June 23, 1912, Turing took an early interest in mathematics. By the age of sixteen, he was studying Einstein's theory of relativity, and it soon became evident that he was a mathematical genius. In fall 1931 he entered King's College at Cambridge to study mathematics. A turning point in his life occurred about the time he graduated: he took a course on the foundations of mathematics from M. H. Newman. During the lectures Newman talked about a list of unsolved problems that the famous mathematician David Hilbert had presented many years earlier. The twenty-third problem on the list intrigued Turing. It was to find a method of establishing the truth or falsity of any statement using formal logic and the standard methods of mathematical proof.[7]

Turing was a long-distance runner at the time and spent a lot of time training. During his long runs, he began thinking about Hilbert's problem and how it could be solved. Interestingly, he didn't approach the problem directly as most others had; after much thought, however, he came up with a unique indirect approach. He cast the problem in terms of computers and computer outputs. Not much was known about computers at the time, so he had to visualize what he called a "universal" computer—a computer that could do anything a human could, but much faster.

The machine he visualized was a relatively simple device that had a printing head which was able to print symbols, one at a time. The symbols were printed on a tape that was segmented, with only one symbol allowed in each segment. His machine had a memory, and information in the memory could be read. The basic actions of the computer were simple and were essentially the same as those of a modern computer. In effect, the machine could compute anything a mathematician with a pencil could calculate, but the machine could do it much faster.[8]

But it wasn't what the machine could do that interested Turing; it was what it couldn't do. He began looking into mathematical problems that were impossible for the machine. It may surprise you that there are such problems. After all, modern computers are incredibly fast and have extremely large memories. They should be capable of doing anything. At least this is what you might think. But surprisingly, they can't work a problem that required more than 100 factorial ($100 \times 99 \times 98 \times \ldots 1$) operations. It is too large for any computer. To see the significance of this, let's translate it into a simple problem: suppose we have a salesman who has 100 cities on his route. A computer could not determine all possible ways of visiting these cities (it is 100 factorial). To see why, we merely have to take a look at how it would have to be calculated. The uncertainty principle of quantum mechanics limits the time for any given operation in a computer to 10^{-43} seconds. This means that any computer is limited to 10^{43} operations per second. Now let's calculate the number of 10^{-43} second intervals in the age of the universe. This gives us our first limit. Yes, you say, but we could have millions of computers hooked together, all working on the same problem at the same time. This is true, but we can easily take it into consideration by looking at the number of particles the universe holds. It is 10^{80}. We could, of course, never make a computer as small as a particle, nevertheless this gives us our final limit. It tells us we could never do more than 10^{150} operations, and 100 factorial is roughly equal to 10^{158}. Surprisingly, there are quite a large number of problems that fit into this category.[9]

Turing became interested in these problems. In particular, he asked himself if a computer could be augmented with a "black box" so that it could solve these problems. He referred to his black box as an *O machine* (where O stands for *oracle*). Along with his imagined computer, which is referred to as a *Turing machine,* the combination is referred to as a *hypermachine*. Using this device, Turing showed that Hilbert's twenty-third problem was impossible. He completed the work in April 1936 when he was twenty-three years old. In later years his machine continued to have a tremendous impact on computer technology.

He took the work to Newman, who was impressed. But oddly enough, Newman had just received the preprint of a paper from a mathematician by the name of Alonzo Church at Princeton University who had also solved Hilbert's problem, but he did it a different way. Newman told Church about Turing's work and suggested that Turing go to Princeton. And in September 1936 Turing arrived in Princeton, and two years later he received his doctorate at the institution.

While at Princeton, Turing met von Neumann, and von Neumann was so impressed with his abilities that he offered him a job as his assistant. Turing, however, declined the offer and returned to England. Shortly after he returned, he began working at Bletchley Park. Details of what he achieved remain secret, but it is known that he helped build a large computer for decoding Enigma. His contribution was so important that after the war he was awarded the Order of the British Empire by the king.

The British had made little progress in decoding the Enigma when Turing arrived at Bletchley, but they had intercepted a lot of cipher material. They weren't sure what to do with it, and they didn't take it as seriously as they should have. When Turing and a few others heard about it, they knew it required immediate action. Going over the heads of their superiors, they wrote directly to Winston Churchill and were given the go-ahead to do anything they could to break the code. And over the next couple of years, the code was broken without the Germans knowing it. This no doubt saved a lot of lives and is believed to have shortened the war.

After the war Turing went to work at the National Physical Laboratory. They were developing a computer called ACE. It was while he was here that he wrote another classic paper on artificial neuron connections and artificial intelligence. Indeed, he developed what is now known as the *Turing Test* for determining whether a computer is "intelligent." His superior, Charles Darwin (grandson of Charles Darwin of evolution fame), dismissed the paper as "childish" and it was not published until fourteen years after Turing's death.

Turing finally got fed up with the delays on ACE and left. His old teacher Newman heard he was available and immediately offered him a job at Manchester University. He worked on a computer design there until his death.

There's no doubt that Turing was a genius, and like many geniuses he was eccentric. While working at Bletchley, he frequently had hay fever, so he began pedaling his bike to work wearing a gas mask. He also converted all his money to silver bars, worried that the Germans were going to invade, and buried them in the woods behind Bletchley Park, but he failed to make an accurate map of their positions and never found them. His life was also, in many ways, tragic. Shortly after going to Manchester, his house was broken into; Turing found out who did it and had him arrested. In the process it came out that Turing was gay, which at that time was a serious offense in England. At his trial he was offered a two-year prison sentence or a one-year rehabilitation program. He took the latter, which included heavy doses of estrogen which caused him to grow breasts, and a hormonal implant in his thigh.

He underwent the treatment but was relieved when it was over. One year later, in June 1954, he committed suicide. He was forty-one.

BACK TO VON NEUMANN

We have already met John von Neumann, but even after ENIAC he continued to play a role in the computer revolution. Born in Hungary in 1903, von Neumann's genius was recognized at an early age. When he was six he could divide two eight-digit numbers in his head with amazing accuracy. He left Hungary after World War I and studied in Germany and Switzerland. He was at Göttingen in the mid-1920s when the quantum mechanical revolution was going on, and he was immediately caught up in it. He showed the equivalence of Heisenberg's and Schrödinger's approach to quantum mechanics, but it came out after Schrödinger's and was hardly noticed. His book *Mathematical Foundations of Quantum Mechanics* was noticed, however, and is now considered to be a masterpiece.[10] He obtained his doctorate from the University of Hamburg in 1926, and in 1930 he emigrated to the United States. After teaching briefly at Princeton University, he took a position along with Einstein at the newly forming Princeton Institute of Advanced Study. At thirty he was the youngest faculty member. He stayed there for the rest of his life.

One of von Neumann's major contributions was his *Theory of Games*. The name might sound frivolous, but it is now considered to be an important branch of mathematics which deals with the best strategy to use in many situations, including business and war. His book *The Theory of Games and Economic Behavior* was published in 1944. Interestingly, it has a chapter titled "Poker and Bluffing," but it's not for the average gambler—it's filled with complicated equations.

His ability to manipulate numbers in his head is legendary.

John von Neumann

One story involves a young engineer who wrote up a detailed mathematical report on a device he had invented. I'm not sure how long the report was, but it was no doubt a few hundred pages and riddled with equations. The engineer caught up with von Neumann one day and gave him the report to read, asking his opinion, thinking von Neumann would take the report with him. Von Neumann quickly read the first few pages and then turned to the last few pages and read them. After only a few minutes he handed the report back to the engineer with the comment, "It won't work." The engineer went back over everything carefully, and several months later he found that von Neumann was right.

After von Neumann left the Moore School of Engineering project, he immediately initiated a project similar to EDVAC at the Princeton Institute. There were no facilities for building such a machine at the institute—it was primarily an academic institution without labs—but von Neumann soon fixed that. And it didn't take him long to find backers for his project; both the Radio Corporation of America (RCA) and the Atomic Energy Commission agreed to support it.

By June 1946 von Neumann had selected his team and begun work. Scientists at RCA promised they would be able to provide a memory unit for the computer. Within a short time, however, problems developed when the first tests were made, and some of the circuits had to be redesigned. Then it became obvious that RCA would not be able to supply the memory unit they had promised. Von Neumann and his team frantically looked around for a replacement. They turned to a magnetic drum, but it was soon found that it would have to spin too fast to keep up with the needs of the computer. Then they heard of a memory unit being developed in England. It would work, but forty would be needed according to von Neumann's design, and they were worried they wouldn't be able to get them in time. Fortunately they got them, and in January 1951 the computer was ready for its final test. Several long calculations were needed for the hydrogen bomb; they would take sixty twenty-four-hour days and would constitute a good test of the speed and stored-program capabilities of the computer. The calculations were completed on time, and the computer was deemed a success. Copies of the machine were soon made at several other laboratories around the United States and overseas.

When completed, von Neumann's computer had 2,300 tubes, which was considerably less than ENIAC. Furthermore, it was much smaller, being only six feet long by eight feet high and two feet wide.[11]

While von Neumann and his team were working on their computer, Eckert and Mauchly had managed to get the support of the Census Bureau and the Bureau of Standards for their project. It was called UNIVAC and was completed in March 1951.

GENERATIONS OF COMPUTERS

All the computers discussed so far are referred to as *first-generation* computers. They were constructed either with mechanical relays and switches or with vacuum tubes. Vacuum tubes were much faster than relays, but they had a limited life and there were problems with reliability. All this changed with the invention of the transistor. And this is where quantum mechanics enters the picture. As we saw earlier, the transistor is a direct product of quantum mechanics in that quantum mechanics allowed us to understand the energy-band structure of semiconductors, and it was through this understanding that we were able to invent the transistor. So it goes without saying that the computer revolution has been made possible because of quantum mechanics.

The transistor soon became the central device of computers, and we entered what is called the *second generation* of computers. Computers became much smaller, faster, and less costly. Furthermore, computer languages were now being developed. All early computers were programmed directly in machine language. With the second generation came assembly languages where abbreviations were used for groups of machine-language instructions. This made things considerably simpler.

Then came the proposed trip to the moon. If we were to land a man on the moon before 1970, transistors and other electronic components had to be made much smaller and more reliable. This led to the *third generation* of computers. Integrated circuits became the norm; they were silicon wafers that held hundreds of tiny transistors along with other electronic components. With these devices computers became even smaller, faster, and more reliable. In addition, new languages such as FORTRAN and COBOL were developed. They were much simpler and easier to use than assembly languages.

But the drive for smaller and faster units continued, and microprocessors were developed. Entire circuits that included thousands of transistors were put on tiny silicon "chips." And this, in turn, gave us *fourth-generation* computers. Small desk-top computers were soon on the market, and supercomputers such as the CRAY were developed.

By the mid-1970s several manufacturers began to sell hobbyist computer kits. Two teenagers, Steve Jobs and Steve Wozniak, saw this as an opportunity. They built their own simple model and began selling it at local stores. They were so successful they soon formed the company Apple Computers. Other companies soon followed. Software began to be developed that allowed one to write and edit reports. Then came commercial spreadsheets and the Internet.

Today, computers of all types surround us. Literally everyone uses them in one way or another whether they know it or not. There are computers in your car. They analyze various functions within the car and in some cases inform you of parts and services that are needed. Indeed, the incredible reliability that we now see in cars is largely due to the computers in them. And computers may play an even larger role in cars of the future. Cars that automatically drive themselves are on the drawing boards. They will be directed by on-board computers and computers along various freeway lanes. Computers are also used extensively in airplanes and in directing air traffic.

Computers are also used in the communications industry, manufacturing, medicine, and agriculture. Computer-controlled robots now perform many routine tasks in building such things as cars, airplanes, appliances, machinery, and electronic devices. Computers are used in weather predicting, monitoring natural resources, space exploration, and undersea exploration. Scientists and engineers use them extensively in their research and development.

Business and commerce rely on computers in many different ways. Libraries now rely on computers to keep track of all their books, and governments could hardly function without them. The list of computer uses goes on. And again, it is quantum mechanics that is the basis of computers.

THE FUTURE

How much further can we go? There's no doubt we are approaching the limit. The *fifth generation* of computers will take us into nanotechology which is getting very close to the scale of the atom, and certainly into the quantum realm—the region where quantum mechanics rules. A nanometer, which is one-billionth of a meter, contains only about five silicon atoms. So nanotechnology is indeed at the border of the quantum realm, and most people agree that's about as far as we will be able to go.

Nanochips are still a thing of the future, but many companies are working hard to develop them. Furthermore, there is now considerable talk about *quantum computers*. At this point they are nothing more than theoretical concepts. Quantum computers will use quantum wave functions and take advantage of various quantum effects such as quantum mechanical interference, and they may be able to solve problems that conventional computers cannot. There's little doubt that quantum mechanics will become increasingly important as attempts are made to produce nanocomputers and quantum computers.

Epilogue

We have come to the end of our story of quantum mechanics and how it has benefited mankind. We have seen how the earliest quantum theory was discovered in 1900 by Planck and how it was used by Bohr to put Rutherford's model of the atom on a solid foundation. The real breakthrough to a complete quantum theory came in 1925 and 1926 when Heisenberg and Schrödinger independently discovered different forms of the theory. Schrödinger later showed they were equivalent. Heisenberg followed up with his uncertainty principle, and Bohr capped the theory off with his principle of complementarity.

We also saw how the theory gave rise to many important devices and developments such as lasers and masers, transistors, superconductors, nuclear physics, and the computer revolution. Indeed, many of the labor-saving devices that we have today are a direct result of the theory. But there are also other areas of science that have benefited from quantum mechanics. One is molecular biology and the study of the DNA molecule. As Crick and Watson showed in

1953, the DNA molecule consists of two interlocking helices, each of which is made up of alternating units of phosphate and sugar. In outward appearance they look like a spiral staircase (fig. 36).

The two sides of the molecule are held together by the base units A(adenine), T(thymine), G(guanine), and C(cytosine). Base unit A bonds only with T (and vice versa) and G bonds only with C (and vice versa). The bonds are relatively weak and can easily be separated. This allows the DNA molecule to replicate, thereby giving an exact copy of itself. It does this by unwinding and exposing the code along its interior (e.g., ATTGCGTA . . .). This is usually referred to as the *code of life*—it codes for everything in the cell, and in the body in general, so it is well protected. Only at replication time is it exposed.

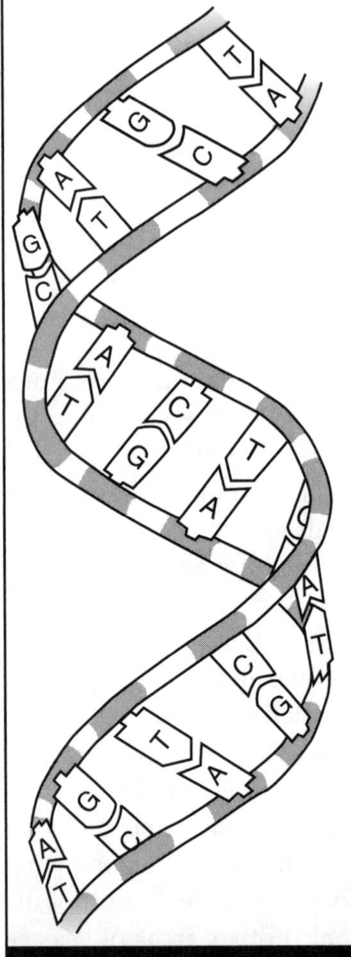

Fig. 36. The simple representation of the DNA molecule showing how A joins with T and G joins with C

If one of the base units was accidentally changed while the code was exposed, when further molecules were produced, there would be an error. Such an error is called a *mutation*. There are several ways such errors can occur. Another molecule, for example, can get in the way at replication time. Also, what is known as "quantum tunneling" can occur. We saw this phenomenon earlier in relation to the nucleus, where an alpha particle quantum tunneled out of it. In the same way as there is a barrier around the nucleus, there is also a barrier holding the base units in DNA in place. A proton, however, can quantum tunnel through this barrier, and if it does, A (assuming the mutation occurred at A) will not bond with T at the next replication, it may bond with G. Similarly, if the mutation occurred at a G, we may not have C and G pairing. The overall result is a mutation. As more and more mutations accumulate, various genetic diseases can occur, one of them being cancer.

As we gain a better understanding of the DNA molecule, quantum mechanics will no doubt play an increasingly important role. It will also no doubt play an increasingly important role in such areas as photosynthesis (the manufacture of organic compounds such as sugars and starches from inorganic substances), bioluminescence (the glow associated with certain biological materials), organic semiconductors, biopolymers, and radiobiology (the study of the effect of radiation on biological materials).

At the other end of the scale, quantum mechanics is also important in our understanding of cosmogony and the universe—particularly the very early universe. In fact, there is considerable speculation that our universe began as a "quantum fluctuation" (a tiny disturbance in the original gas). Ed Tryon of Hunter College, New York, put forward a theory based on this idea in 1973. Furthermore, it is believed that quantum phenomena played an important role shortly after the universe was born. The huge cloud that went out in the explosion had to break up and produce the galaxies we now see throughout the universe. Scientists believe that quantum fluctuations very early on created the breakup.

Black holes, the small bizarre objects from which nothing—not even light—can escape, also appear to have a quantum mechanical connection. This is strange in that we know that black holes are explained by Einstein's theory of relativity. Stephen Hawking of Cambridge University, however, has shown that they give off both radiation and short-lived particles. Of particular importance he showed that this radiation obeys Planck's law, in other words, the law that gave us quantum theory. This is a significant result. No connection has ever been found between the two great theories of physics, namely, general relativity and quantum mechanics. They appear to be entirely separate theories. Numerous attempts have been made to quantize general relativity, thereby bringing the two theories together, but none has been successful. Such a theory is needed in relation to the very early universe; in fact, without it we really can't understand the earliest events of creation. Hawking showed that there is a "slight crack" in the shield between the two theories. Much more needs to be done before we can understand the connection thoroughly, but it is at least a first step.

Quantum mechanics is truly a discovery that has changed the world, and with its application to cosmology, black holes, galaxies, and other astronomical phenomena, it has also changed the way we look at the universe.

Notes

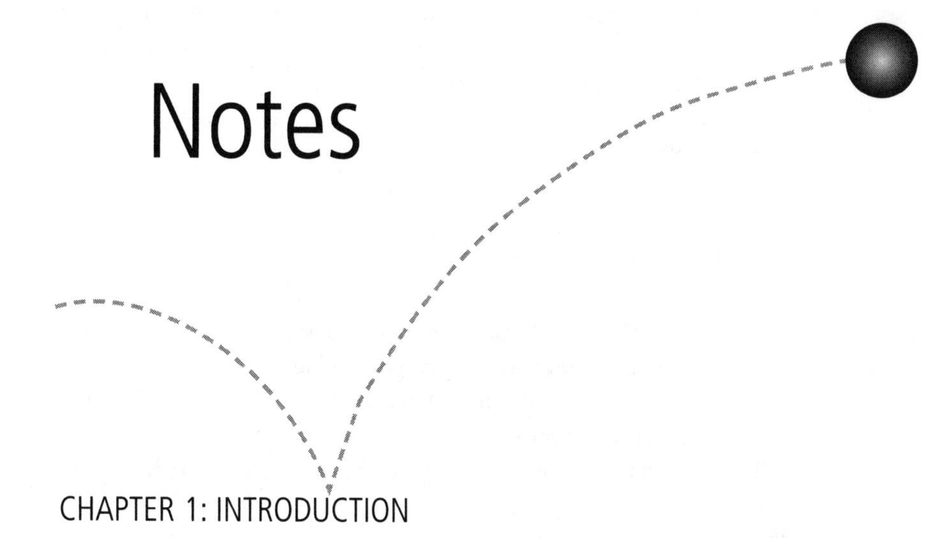

CHAPTER 1: INTRODUCTION

1. The spectrum of hydrogen is a series of bright (or dark) lines against a dark background that occurs when the light from glowing hydrogen is passed through an instrument called a spectroscope. The spectroscope spreads out the various wavelengths.

2. A differential equation is a type of equation that occurs in calculus. It tells us how a system changes; in particular, it can tell us how the system evolves in time.

3. Charge density would give a measure of the charge from point to point.

4. CERN is the large high-energy laboratory in Europe. It is located near Geneva on the Switzerland-France border. It is a consortium of European nations.

5. Photons are particles of energy. Light, for example, consists of photons.

CHAPTER 2: EARLY IDEAS

1. 10^{23} is 1 followed by 23 zeros.

2. Albrecht Fölsing, *Albert Einstein: A Biography* (New York: Viking, 1997), p. 126.

3. David Wilson, *Rutherford: Simple Genius* (Cambridge: MIT Press, 1983), p. 109.

4. Ibid., p. 125.

5. Ibid., p. 62.

6. Ibid., p. 291.

CHAPTER 3: THE LUCKY GUESS

1. Jagdish Mehra and Helmut Rechenberg, *The Historical Development of Quantum Theory*, vol. 1 (New York: Springer-Verlag, 1982), p. 43.

2. Max Jammer, *The Conceptual Development of Quantum Mechanics* (New York: McGraw-Hill, 1966), p. 22.

3. Albrecht Fölsing, *Albert Einstein: A Biography* (New York: Viking, 1997), p. 84.

4. Ibid., p. 145.

5. John Stachel, *Einstein's Miraculous Year* (Princeton: Princeton University Press, 1998), p. 177.

6. Fölsing, *Albert Einstein*, p. 147.

7. The specific heat of a substance is defined as the quantity of heat required to raise the temperature of one gram of that substance by one degree centigrade.

CHAPTER 4: THE BOHR ATOM

1. S. Rozental, *Niels Bohr: His Life and Work as Seen by His Friends and Colleagues* (Amsterdam: North-Holland, 1967), p. 15.

2. Ibid., p. 34.

3. Ibid., p. 43.

4. Ibid., p. 46.

5. Jagdish Mehra and Helmut Rechenberg, *The Historical Development of Quantum Theory*, vol. 1 (New York: Springer-Verlag, 1982), p. 188.

6. Ibid., p. 193.

7. Ibid., p. 201.

8. This refers to Einstein's special theory of relativity that he formulated in 1915. It deals mainly with the properties of space and time.

9. Rozental, *Niels Bohr*, p. 67.
10. Robert March, *Physics for Poets* (New York: McGraw-Hill, 1978), p. 204.
11. A single crystal has a lattice that has a periodic structure throughout.

CHAPTER 5: HEISENBERG'S ARRAYS

1. Jagdish Mehra and Helmut Rechenberg, *The Historical Development of Quantum Theory*, vol. 2 (New York: Springer-Verlag, 1982), p. 208.

2. David Cassidy, *Uncertainty: The Life and Science of Werner Heisenberg* (New York: Freeman and Company, 1992), p. 100.

3. Ibid., p. 102.

4. Mehra and Rechenberg, *The Historical Development of Quantum Theory*, vol. 2, p. 30.

5. Cassidy, *Uncertainty*, p. 129.

6. Ibid., p. 142.

7. An interferometer is a device in which beams of light are brought together to produce an interference pattern. In some cases the beams interfere constructively and reinforce one another, in others they interfere destructively and neutralize one another.

8. Cassidy, *Uncertainty*, p. 152.

9. Ibid., p. 153.

10. Mehra and Rechenberg, *The Historical Development of Quantum Theory*, vol. 2, p. 138.

11. Ibid., p. 153.

12. Ibid., p. 248.

13. Ibid., p. 249.

14. Cassidy, *Uncertainty*, p. 236.

CHAPTER 6: SCHRÖDINGER'S WAVE EQUATION

1. Walter Moore, *Schrödinger: Life and Thought* (Cambridge: Cambridge University Press, 1989), p. 17.

2. Ibid., p. 23.

3. Ibid., p. 46.

4. Ibid., p. 54.

5. Jagdish Mehra and Helmut Rechenberg, *The Historical Development of Quantum Theory*, vol. 5 (New York: Springer-Verlag, 1982), p. 184.

6. Moore, *Schrödinger,* p. 91.

7. Mehra and Rechenberg, *The Historical Development of Quantum Theory,* vol. 5, p. 326.

8. Moore, *Schrödinger,* p. 141.

9. I have referred to Hermann Weyl's book *Space-Time-Matter* several times. It can be obtained from Dover Publications in New York.

10. Mehra and Rechenberg, *The Historical Development of Quantum Theory,* vol. 5, p. 412.

11. Ibid., p. 420.

12. Ibid., p. 421.

13. Moore, *Schrödinger,* p. 196.

14. Ibid., p. 209.

CHAPTER 7: WHAT DID IT ALL MEAN?

1. Jagdish Mehra and Helmut Rechenberg, *The Historical Development of Quantum Theory,* vol. 5 (New York: Springer-Verlag, 1982), p. 636.

2. Ibid., p. 617.

3. Walter Moore, *Schrödinger: Life and Thought* (Cambridge: Cambridge University Press, 1989), p. 221.

4. David Cassidy, *Uncertainty: The Life and Science of Werner Heisenberg* (New York: Freeman and Company, 1992), p. 222.

5. Mehra and Rechenberg, *The Historical Development of Quantum Theory,* vol. 5, p. 639.

6. Ibid., p. 640.

7. The words *eigenfunction* and *eigenvalue* come from the German language. Eigenfunctions are the solutions of differential equations such as Schrödinger's equation. Each corresponds to a different state of the system. Eigenvalues are numbers. For example, they could correspond to the energy levels within a given state.

8. Robert March, *Physics for Poets* (New York: McGraw-Hill, 1978), p. 210.

9. Ibid., p. 215.

10. Moore, *Schrödinger,* p. 226.

11. Ibid., p. 228.

12. Moore, *Schrödinger,* p. 230.

13. Ibid., p. 239.

CHAPTER 8: UNCERTAINTY

1. S. Rozental, *Niels Bohr: His Life and Work as Seen by His Friends and Colleagues* (Amsterdam: North-Holland, 1967), p. 105.

2. Robert March, *Physics for Poets* (New York: McGraw-Hill, 1978), p. 216.

3. Rozental, *Niels Bohr,* p. 106.

4. David Cassidy, *Uncertainty: The Life and Science of Werner Heisenberg* (New York: Freeman and Company, 1992), p. 242.

5. Fred Alan Wolf, *Taking the Quantum Leap* (San Francisco: Harper and Row, 1981), p. 130.

6. Ibid., p. 122.

7. March, *Physics for Poets,* p. 221.

8. Cassidy, *Uncertainty,* p. 248.

9. Walter Moore, *Schrödinger: Life and Thought* (Cambridge: Cambridge University Press, 1989), p. 239.

10. Rozental, *Niels Bohr,* p. 197.

11. Ibid., p. 108.

CHAPTER 9: EINSTEIN'S OBJECTIONS AND QUANTUM WEIRDNESS

1. John Gribbin, *In Search of Schrödinger's Cat* (New York: Bantam Books, 1984), p. 178.

2. Brian Greene, *The Elegant Universe* (New York: Norton, 1999), p. 96.

3. Heinz Pagels, *The Cosmic Code: Quantum Physics as the Language of Nature* (New York: Simon and Schuster, 1982), p. 160.

4. Arthur Fine, *The Shaky Game: Einstein's Realism and the Quantum Theory* (Chicago: University of Chicago Press, 1986), p. 29.

5. Ibid., p. 32.

6. Gribbin, *In Search of Schrödinger's Cat,* p. 216

7. S. Rozental, *Niels Bohr: His Life and Work as Seen by His Friends and Colleagues* (Amsterdam: North-Holland, 1967), p. 128.

8. Walter Moore, *Schrödinger: Life and Thought* (Cambridge: Cambridge University Press, 1989), p. 304.

9. Pagels, *Cosmic Code,* p. 148.

10. Fred Alan Wolf, *Taking the Quantum Leap* (San Francisco: Harper and Row, 1981), p. 205.

11. Gribbin, *In Search of Schrödinger's Cat,* p. 227.

CHAPTER 10: EXTENDING THE THEORY

1. Barry Parker, *Search for a Supertheory* (New York: Plenum, 1987), p. 79.

2. Gary Zukav, *The Dancing Wu Li Masters* (New York: Morrow, 1979), p. 240.

3. Parker, *Search for a Supertheory,* p. 82.

4. Ibid., p. 84.

5. Feynman has written several books about his life and his work, including *Surely You're Joking, Mr. Feynman* (New York: Norton, 1985) and others.

6. Many of Feynman's "funny little diagrams" and their explanation can be found in Gary Zukav's book, *The Dancing Wu Li Masters* (New York: Morrow, 1979) and in a book by J. E. Dodd, *The Ideas of Particle Physics* (Cambridge: Cambridge University Press, 1984).

7. Dodd, *The Ideas of Particle Physics,* p. 37.

8. One of the best books that follow up on the latest attempts to formulate an extension of this theory and a unified theory of all physics is Brian Greene's book, *The Elegant Universe* (New York: Norton, 1999).

CHAPTER 11: MODERN DEVELOPMENTS: LASERS AND MASERS

1. Later editions of this book are referred to as *Relativity: The Special and General Theory* (New York: Crown, 1952).

2. M. Bertolotti, *Lasers and Masers: An Historical Approach* (Bristol: Adam Hilger, 1983), p. 7.

3. Ibid., p. 20.

4. Ibid., p. 71.

5. Ibid., p. 76.

6. Charles Pike, *Lasers and Masers* (New York: Bobbs-Merrill, 1967), p. 75.

7. Bertolotti, *Lasers and Masers,* p. 75.

8. Ibid., p. 84.

9. Ibid., p. 75. George Trigg's book *Landmark Experiments in Twentieth Century Physics* (New York: Crane, Russak, 1975) also gives a good overview of this experiment.

10. Bertolotti, *Lasers and Masers,* p. 110.

11. Ibid., p. 120.

12. A semiconductor is a material between a conductor and an insulator.

13. Pike, *Lasers and Masers*, p. 105.

14. Bertolotti, *Lasers and Masers*, p. 166.

15. Many applications of lasers and masers can be found in Stan Gibilisco, *Understanding Lasers* (Blue Ridge Summit: Tab Books, 1989) and in Allan Lytel, *ABC's of Lasers and Masers* (New York: Bobbs-Merrill, 1968).

CHAPTER 12: TRANSISTORS AND SUPERCONDUCTORS

1. Linda Edwards-Shea, *The Essence of Solid State Electronics* (New York: Prentice-Hall, 1996), p. 39.

2. Alan Holden, *The Nature of Solids* (New York: Columbia University Press, 1965), p. 200.

3. *Nobel Lectures in Physics: 1942–1962* (Amsterdam: Elsevier, 1964), p. 318.

4. George Trigg, *Landmark Experiments in Twentieth Century Physics* (New York: Crane, Russak, 1975), p. 149.

5. *Nobel Lectures in Physics*, p. 344.

6. Bruce Schecter, *The Path of No Resistance: The Story of the Revolution in Superconductivity* (New York: Simon and Schuster, 1989), p. 40.

7. Ibid., p. 46.

8. Ibid., p. 129.

9. Ibid., p. 64.

10. Robert Hazen, *The Race for the Superconductor* (New York: Summit, 1989)

11. Per Fridtjof Dahl, *Superconductivity* (New York: American Institute of Physics, 1992), p. 286.

CHAPTER 13: THE NUCLEAR AGE

1. Richard Rhodes, *The Making of the Atomic Bomb* (New York: Simon and Schuster, 1986), p. 163. This is a particularly good reference for this chapter.

2. David Cassidy, *Uncertainty: The Life and Science of Werner Heisenberg* (New York: Freeman and Company, 1992), p. 293.

3. David Dietz, *Atomic Science, Bombs, and Power* (New York: Dodd, Mead, and Company, 1955), p. 125.

4. George Gamow has written several books about his life and work. One of the best is *My World Line* (New York: Viking, 1970).

5. Barry Parker, *Search for a Supertheory* (New York: Plenum, 1987), p. 127.

6. Rhodes, *The Making of the Atomic Bomb*, p. 242.

7. Ibid., p. 250.

8. Ibid., p. 256.

9. S. Rozental, *Niels Bohr: His Life and Work as Seen by His Friends and Colleagues* (Amsterdam: North-Holland, 1967), p. 145.

10. William Lanouette, "The Odd Couple and the Bomb," *Scientific American* (November 2000): 104.

11. Rhodes, *The Making of the Atomic Bomb*, p. 295.

12. Ibid., p. 304.

13. Dietz, *Atomic Science, Bombs, and Power*, p. 156.

14. Ibid., p. 158.

15. Cassidy, *Uncertainty*, p. 437.

16. Rhodes, *The Making of the Atomic Bomb*, p. 448.

17. Ibid., p. 459.

18. Ibid., p. 674.

CHAPTER 14: THE COMPUTER REVOLUTION

1. Michael Williams, *A History of Computing Technology* (Englewood Cliffs: Prentice-Hall, 1985), p. 187.

2. Ibid., p. 168.

3. Isaac Asimov, *Biographical Encylopedia of Science and Technology* (New York: Doubleday, 1982), p. 323.

4. Williams, *A History of Computing Technology*, p. 214.

5. Ibid., p. 271.

6. Jeremy Bernstein, *Cranks, Quarks, and the Cosmos: Writings on Science* (New York: Basic Books, 1993), p. 92; A. Hodges, *Alan Turing: The Enigma* (New York: Simon and Schuster, 1983).

7. John Hopcroft, "Turing Machines," *Scientific American* (May 1984): 86.

8. Jack Copeland and Diane Proudfoot, "Alan Turing's Forgotten Ideas in Computer Science," *Scientific American* (April 1999): 98.

9. William May, *Edges of Reality* (New York: Insight, 1996), p. 25.

10. A later edition of this book is available from Princeton University Press.

11. Williams, *A History of Computing Technology*, p. 359.

Glossary

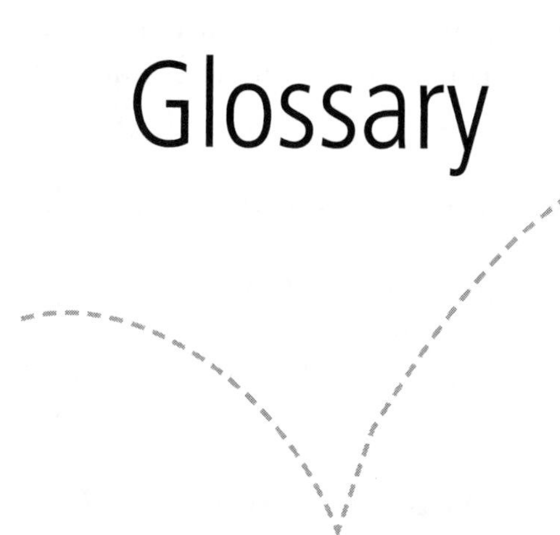

Alpha ray. A helium nucleus.

Anode. Positive electrode.

Antiparticle. Corresponding to every particle there is an antiparticle. When a particle and an antiparticle meet, they annihilate with the release of energy.

Beta ray. Beam of high-speed electrons.

Blackbody radiation. Radiation emitted by a hypothetical body that absorbs and emits all radiation that falls on it.

Canal rays. Positively charged rays (an early name).

Cathode. Negative electrode of a battery.

Cathode rays. A negative current of particles. Electrons.

CAUSALITY. Associated with the idea that everything has to have a cause.

CHAIN REACTION. A reaction in which the fission of one atomic nucleus gives off enough neutrons to cause the fission of another nucleus.

CHANNEL RAYS. Older name for beam of negatively charged particles.

CHIP. A small silicon chip containing many electronic components.

CLASSICAL THEORY. Newtonian theory or any theory that does not involve quantum mechanics.

COMMUTATIVE. Two variables a and b commute when $a \times b = b \times a$.

COMPLEMENTARITY. Bohr's theory that the wave-particle aspects of light are complementary but exclusive.

COMPLEX NUMBER. A generalization of the real number system that includes numbers with a real and an imaginary part.

CONDUCTION BAND. A band of energy levels. If an electron is in the conduction band, it moves within the lattice.

CONDUCTOR. A material that conducts electrons.

DETERMINISM. The idea that everything in the universe is determined by specific laws.

DIAMAGNETISM. Material that is weakly repelled when brought near a magnet.

DIFFERENTIAL EQUATION. A calculus equation that gives the evolution of change in a system.

EIGENFUNCTION. The solution of a differential equation.

EIGENVALUE. Numbers that are given by a differential equation when it is solved.

ELECTRIC FIELD. Field around an electric charge.

ELECTRODE. Terminal of a battery or source of voltage.

ELECTROMAGNETIC WAVE. A wave given off from vibrating electric charges.

ELECTRON. The lightest massive elementary particle. Has negative charge.

ENERGY STATE. A measure of the amount of energy a particle or system has.

EXCHANGE PARTICLE. Particles that are passed back and forth creating a force. The exchange particle of the electromagnetic field is the photon.

FACTORIAL. The product of all the positive integers from 1 to a given number. Ten factorial is $10 \times 9 \times 8 \times \ldots 1$.

FERROELECTRIC. Electric material corresponding to magnets.

FISSION. To break in half. The uranium nucleus fissions when it becomes unstable.

FLUORESCENCE. Giving off light as a result of absorbing radiation of differing wavelengths.

FREQUENCY. The number of vibrations per second.

GAMMA RAY. Highly energetic electromagnetic rays.

General relativity. Einstein's theory of gravity.

Half-life. Time for half of a radioactive sample to decay to a lighter element.

Harmonic oscillator. An oscillator that vibrates with harmonic or concordant motion.

Hidden variable. An underlying or unseen variable in a theory. A "hidden theory" within a theory.

Hole. Place where a particle is missing. Can conduct just as charged particles conduct.

Hydrodynamics. The study of the dynamics of fluid flow.

Interference. The ability of two or more waves to interact causing cancellation or reinforcement.

Ion. An atom or particle with a positive or negative charge.

Isotope. Nuclei containing the same number of protons but a different number of neutrons. Most elements can exist in several isotopic forms.

Laser. Stands for "Light Amplification by the Stimulated Emission of Radiation."

Luminescence. The giving off of light, or the natural release of photons.

Maser. Stands for "Microwave Amplification by the Stimulated Emission of Radiation."

Matrix. An array of numbers used in mathematics.

Moderator. Material for slowing down neutrons.

Momentum. Mass times velocity.

Neutron. Neutral particle of the nucleus.

Nuclear reaction. A reaction involving particles and nuclei.

Nuclear reactor. An installation in which nuclei react with protons and neutrons to produce energy.

Nucleus. At the center of the atom. Positively charged and contains most of the mass.

Paramagnetism. A weakened form of the usual magnetism we are familiar with.

Photoelectric effect. The effect that occurs when light is shone on a metal. Particles are released above a certain frequency.

Photon. Particle of light.

Planck's constant. Usually designated by *h*. The fundamental constant of quantum mechanics.

Population inversion. When a high energy level has more particles in it than a lower energy level.

Positron. The antiparticle of the electron.

Proton. Heavy positively charged particle of the nucleus.

Quanta. A small discrete amount of energy that is absorbed or emitted in particle interactions.

Quantum electrodynamics. Quantum theory of the electromagnetic interactions.

Quantum jump. A jump between quantum orbits or levels.

Quantum number. Numbers that specify quantized physical quantities such as spin or momentum.

Quantum tunnel. "Sneaking" through a barrier under the guise of the uncertainty principle.

Radioactivity. The natural emission of alpha, beta, and gamma rays from unstable nuclei.

Radioisotope. Radioactive isotopes of various elements.

Rectification. The flow of current in one direction only.

Relativity. Refers to Einstein's theory of relativity.

Renormalization. A process in which apparent infinities in various interaction calculations are eliminated.

Resonant cavity. A region in a laser or maser where reflection takes place from the walls creating a buildup of photons.

Schrödinger equation. Quantum mechanical equation governing a system.

Semiconductor. A material that conducts electrons, but does not do it as well as a conductor.

Specific heat. The heat energy absorbed per unit mass, per degree rise in temperature.

Spectroscopy. The study of spectral lines.

Spectrum. Bright or dark lines that occur when light is passed through a spectroscope.

Spin. Refers to a property of a particle that is similar to the spin of a spinning top.

Spontaneous Transformation. The change that occurs when alpha, beta, and gamma rays are released. Occurs naturally.

Stark effect. The splitting of spectral lines in an electric field.

Statistical mechanics. The study of large numbers of particles such as the molecules of a gas, using statistics and probability.

Stimulated emission. The emission of radiation created by stimulation. Associated with electrons falling to lower energy levels.

Strong interactions. Reactions between particles in the nucleus. Very short-ranged.

Theory of numbers. A branch of mathematics dealing with the properties of numbers.

Thermodynamics. The study of the dynamics of heat.

Uncertainty. Refers to the lack of certainty associated with the uncertainty principle.

Virtual photon. Very short-lived photons that pass between particles giving rise to a force.

Viscosity. A measure of the "stickiness" or internal friction of a fluid.

Wave function. A function that represents a system.

Wavelength. Distance between equivalent points along a wave.

Wave mechanics. Refers to quantum mechanics in its wave formulation, or Schrödinger's theory.

Wave packet. A superposition of waves which gives a simple representation of a particle.

Weak interactions. Interactions associated with the nucleus. Much weaker than strong interactions.

Wilson cloud chamber. A device that allows one to see the track of particles that pass through it.

X ray. Energetic electromagnetic radiation between gamma rays and UV.

Zeeman effect. The splitting of spectral lines due to the presence of a magnetic field.

Bibliography

CHAPTER 2: EARLY IDEAS

Gribbin, John. *In Search of Schrödinger's Cat.* New York: Bantam Books, 1984.

Guillemin, Victor. *The Story of Quantum Mechanics.* New York: Scribner's, 1968.

Jammer, Max. *The Conceptual Development of Quantum Mechanics.* New York: McGraw-Hill, 1966.

March, Robert. *Physics for Poets.* New York: McGraw-Hill, 1978.

Wilson, David. *Rutherford.* Cambridge: MIT Press, 1983.

CHAPTER 3: THE LUCKY GUESS

Gribbin, John. *In Search of Schrödinger's Cat.* New York: Bantam Books, 1984.

Heilbron, J. L. *The Dilemmas of an Upright Man.* Berkeley: University of California Press, 1986.

March, Robert. *Physics for Poets.* New York: McGraw-Hill, 1978.

Mehra, Jagdish, and Helmut Rechenberg. *The Historical Development of Quantum Theory.* New York: Springer-Verlag, 1982.
Planck, Max. *The Philosophy of Physics.* New York: Norton, 1963.
Wolf, Fred. *Taking the Quantum Leap.* San Francisco: Harper and Row, 1981.

CHAPTER 4: THE BOHR ATOM

d'Abro, A. *The Rise of the New Physics.* New York: Dover, 1951.
Jammer, Max. *The Conceptual Development of Quantum Mechanics.* New York: McGraw-Hill, 1966.
Mehra, Jagdish, and Helmut Rechenberg. *The Historical Development of Quantum Theory.* New York: Springer-Verlag, 1966.
Pagels, Heinz. *The Cosmic Code.* New York: Simon and Schuster, 1982.
Rozental, S., ed. *Niels Bohr: His Life and Works as Seen by his Friends and Colleagues.* Amsterdam: North-Holland, 1967.
Wolf, Fred. *Taking the Quantum Leap.* San Francisco: Harper and Row, 1981
Zukav, Gary. *The Dancing Wu Li Masters.* New York: Morrow, 1979.

CHAPTER 5: HEISENBERG'S ARRAYS

Cassidy, David. *Uncertainty: The Life and Science of Werner Heisenberg.* New York: Freeman and Company, 1992.
Guillemin, Victor. *The Story of Quantum Mechanics.* New York: Scribner's, 1968.
Heisenberg, Werner. *Physics and Beyond.* New York: Harper and Row, 1971.
Jammer, Max. *The Conceptual Development of Quantum Mechanics.* New York: McGraw-Hill, 1966.
Mehra, Jagdish, and Helmut Rechenberg. *The Historical Development of Quantum Mechanics.* New York: Springer-Verlag, 1982.
Wolf, Fred. *Taking the Quantum Leap.* San Francisco: Harper and Row, 1981.

CHAPTER 6: SCHRÖDINGER'S WAVE EQUATION

Jammer, Max. *The Conceptual Development of Quantum Mechanics.* New York: McGraw-Hill, 1966.

Mehra, Jagdish, and Helmut Rechenberg. *The Historical Development of Quantum Theory.* New York: Springer-Verlag, 1982.

Moore, Walter. *Schrödinger: Life and Thought.* Cambridge: Cambridge University Press, 1987.

Schrödinger, Erwin. *Collected Papers and Wave Mechanics.* New York: Chelsea, 1978.

———. *Lectures on Wave Mechanics.* New York: Philosophical Library, 1967.

CHAPTER 7: WHAT DID IT ALL MEAN?

Capra, Fritjof. *The Tao of Physics.* New York: Bantam, 1980.

Mehra, Jagdish, and Helmut Rechenberg. *The Historical Development of Quantum Theory.* New York: Springer-Verlag, 1982.

Pagels, Heinz. *The Cosmic Code.* New York: Simon and Schuster, 1982.

Wolf, Fred. *Taking the Quantum Leap.* San Francisco: Harper and Row, 1981.

Zukav, Gary. *The Dancing Wu Li Masters.* New York: Morrow, 1979.

CHAPTER 8: UNCERTAINTY

Cassidy, David. *Uncertainty: The Life and Science of Werner Heisenberg.* New York: Freeman and Company, 1992.

Jammer, Max. *The Conceptual Development of Quantum Mechanics.* New York: McGraw-Hill, 1966.

Mehra, Jagdish, and Helmut Rechenberg. *The Historical Development of Quantum Theory.* New York: Springer-Verlag, 1982.

CHAPTER 9: EINSTEIN'S OBJECTIONS AND QUANTUM WEIRDNESS

Fine, Arthur, *The Shaky Game.* Chicago: University of Chicago Press, 1986.

Greene, Brian. *The Elegant Universe.* New York: Norton, 1999.

Gribbin, John. *In Search of Schrödinger's Cat.* New York: Bantam, 1984.
Pagels, Heinz. *The Cosmic Code*. New York: Simon and Schuster, 1982.

CHAPTER 10: EXTENDING THE THEORY

Dodd, J. E. *The Ideas of Particle Physics.* Cambridge: Cambridge University Press, 1984.
Parker, Barry. *Search for a Supertheory.* New York: Plenum, 1987.

CHAPTER 11: MODERN DEVELOPMENTS: LASERS AND MASERS

Bertolotti, Mario. *Masers and Lasers.* Bristol: Alger Hiss, 1983.
Gibilisco, Stan. *Understanding Lasers.* Blue Ridge Summit: Tab, 1989.
Lytel, Allan. *ABC's of Lasers and Masers.* New York: Bobbs-Merrill, 1968.
Pike, Charles. *Lasers and Masers.* New York: Bobbs-Merrill, 1968.

CHAPTER 12: TRANSISTORS AND SUPERCONDUCTORS

Dahl, P. F. *Superconductivity.* New York: American Institute of Physics, 1992.
Edwards-Shea, L. *The Essence of Solid State Electronics*. New York: Prentice-Hall, 1996.
Hazen, Robert. *The Breakthrough.* New York: Simon and Schuster, 1988.
Holden, Alan. *The Nature of Solids.* New York: Columbia University Press, 1965.
Schecter, Bruce. *The Path of No Resistance*. New York: Simon and Schuster, 1989.
Trigg, George. *Landmark Experiments in Twentieth Century Physics.* New York: Crane, Russak and Company, 1975.

CHAPTER 13: THE NUCLEAR AGE

Dietz, David. *Atomic Science, Bombs, and Power.* New York: Dodd, Mead and Company, 1954.

Lanoutte, William. "The Odd Couple and the Bomb." *Scientific American* (November 2000): 104.

Rhodes, Richard. *The Making of the Atomic Bomb.* New York: Simon and Schuster, 1986.

Seaborg, Glenn. *Nuclear Milestones.* Oak Ridge: Atomic Energy Commission, 1971.

Wendt, Gerald. *Atomic Energy and the Hydrogen Bomb.* New York: Medill McBride, 1950.

CHAPTER 14: THE COMPUTER REVOLUTION

Berkeley, E. C. *Great Brains, or Machines that Think*. New York: Science Edition, 1961.

Bernstein, Jeremy. *Cranks, Quarks, and the Cosmos.* New York: Basic Books, 1993.

Deutsch, David. *The Fabric of Reality.* New York: Penguin, 1997.

Hodges, Andrew. *Alan Turing: The Enigma*. New York: Simon and Schuster, 1983.

May, W. D. *Edges of Reality.* New York: Insight Books, 1996.

Williams, M. R. *A History of Computing Technology.* Englewood Cliffs: Prentice-Hall, 1985.

EPILOGUE

Parker, Barry. *Cosmic Time Travel.* New York: Plenum, 1991.

———. *The Vindication of the Big Bang*. New York: Plenum, 1993.

Watson, J. *The Double Helix.* New York: Dutton, 1960.

Index